WITHDRAWN BY THE
UNIVERSITY OF MICHIGAN

North American sturgeons:
biology and aquaculture potential

Developments in environmental biology of fishes 6

Series Editor
EUGENE K. BALON

North American sturgeons: biology and aquaculture potential

Papers from a symposium on the biology and management of sturgeon, held during the 113th annual meeting of the American Fisheries Society at Milwaukee, Wisconsin, U.S.A., August 16–20, 1983

Edited by
FREDERICK P. BINKOWSKI & SERGEI I. DOROSHOV

Reprinted from *Environmental biology of fishes* 14 (1), 1985 with addition of six more papers from the symposium and an epilogue on sturgeon culture

1985 **DR W. JUNK PUBLISHERS**
a member of the KLUWER ACADEMIC PUBLISHERS GROUP
DORDRECHT / BOSTON / LANCASTER

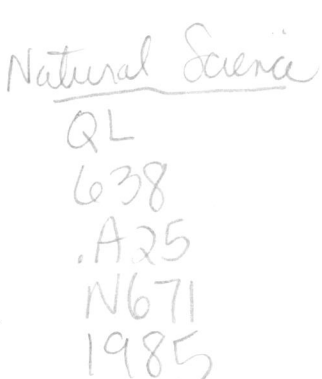

Natural Science
QL
638
.A25
N671
1985

Distributors

for the United States and Canada:
Kluwer Boston, Inc., 190 Old Derby Street,
Hingham, MA 02043, USA
for the UK and Ireland:
Kluwer Academic Publishers, MTP Press Limited, Falcon
House, Queen Square, Lancaster LA1 1RN, England
for all other countries:
Kluwer Academic Publishers Group, Distribution Center,
P.O. Box 322, 3300 AH Dordrecht, The Netherlands

Library of Congress Cataloging in Publication Data

```
North American sturgeons.

   (Developments in environmental biology of fishes ; 6)
   "Reprinted from Environmental biology of fishes,
14 (1), 1985 with addition of six more papers from the
symposium and an epilogue on sturgeon culture."
   Includes index.
   1. Sturgeons--Congresses.  2. Fish-culture--
United States--Congresses.  3. Fishery management--
United States--Congresses.  I. Binkowski, Frederick P.
II. Doroshov, Sergei I.  III. American Fisheries Society.
Meeting (113th : 1983 : Milwaukee, Wis.)  IV. Series.
QL638.A25N67  1985       639.3'744       85-23095
```

ISBN 90-6193-539-3 (this volume)
ISBN 90-6193-896-1 (series)

Cover design: Max Velthuijs

Copyright

© 1985 by Dr W. Junk Publishers, Dordrecht

All rights reserved. No part of this publication may be reproduced, stored in a retrieval system, or transmitted in any form or by any means, mechanical, photocopying, recording, or otherwise, without the prior written permission of the publisher,
Dr W. Junk Publishers, P.O. Box 163, 3300 RD Dordrecht, The Netherlands.

PRINTED IN THE NETHERLANDS

Contents

Preface, by F.P. Binkowski & S.I. Doroshov	7
Introduction, by R.A. Ragotzkie	9
Gamete interaction in the white sturgeon *Acipenser transmontanus*: a morphological and physiological review, by G.N. Cherr & W.H. Clark, Jr.	11
Osmoregulation in juvenile and adult white sturgeon, *Acipenser transmontanus*, by M. McEnroe & J.J. Cech, Jr.	23
Digestive and feeding characteristics of the chondrosteans, by R.K. Buddington & J.P. Christofferson	31
Effect of temperature on early development of white and lake sturgeon, *Acipenser transmontanus* and *A. fulvescens*, by Y.L. Wang, F.P. Binkowski & S.I. Doroshov	43
Distribution, biology and hybridization of *Scarphirhynchus albus* and *S. platorynchus* in the Missouri and Mississippi rivers, by D.M. Carlson, W.L. Pflieger, L. Trial & P.S. Haverland	51
The fishery, biology, and management of Atlantic sturgeon, *Acipenser oxyrhynchus*, in North America, by T.I.J. Smith	61
The lake sturgeon, *Acipenser fulvescens*, in the Menominee River, Wisconsin-Michigan, by T.F. Thuemler	73
Artificial spawning and rearing of lake sturgeon, *Acipenser fulvescens*, in Wild Rose State Fish Hatchery, Wisconsin, 1982–1983, by D.G. Czeskleba, S. AveLallemant & T.F. Thuemler	79
Oocyte maturation in white sturgeon, *Acipenser transmontanus*: some mechanisms and applications, by P.B. Lutes	87
The effect of food deprivation on the plasma free amino acid levels of sturgeon, *Acipenser transmontanus*, by R.L. Swallow	93
Evaluation of morphometric characters used in taxonomic separation of Gulf of Mexico sturgeon, *Acipenser oxyrhynchus desotoi*, by C.M. Wooley	97
Preliminary description of the genetic structure of white sturgeon, *Acipenser transmontanus*, in the Pacific Northwest, by D.M. Bartley, G.A.E. Gall & B. Bentley	105
Habitat use and behavior of pre-spawning and spawning shortnose sturgeon, *Acipenser brevirostrum*, in the Connecticut River, by J. Buckley & B. Kynard	111
Status, life history, and management of Columbia River white sturgeon, *Acipenser transmontanus*, by J.L. Galbreath	119
Status of white sturgeon, *Acipenser transmontanus*, in Idaho, by T.G. Cochnauer, J.R. Lukens & F.E. Partridge	127
Management of the lake sturgeon, *Acipenser fulvescens*, population in the Lake Winnebago system, Wisconsin, by D.J. Folz & L.S. Meyers	135
Epilogue: a perspective on sturgeon culture, by S.I. Doroshov & F.P. Binkowski	147
Species and subject index	153

F.P. Binkowski & S.I. Doroshov (ed.), North American sturgeons. ISBN 90-6193-539-3
© 1985, Dr W. Junk Publishers, Dordrecht. Printed in the Netherlands. Developments in EBF 6.

Preface

Sturgeon are an important group of commercial and recreational fishes. These ancient fish evolved over 200 million years ago. Approximately 25 species have survived to the present time and all are found in the northern hemisphere.

In some regions these fish are prized for their commercial value, including flesh and caviar through both natural harvest and artificial culture. Centuries ago during the colonization of North America these valuable fishes were once thought of as trash fish. Later man appreciated the value of these fish, as a result many stocks were over harvested. Following the decline of populations in North America during the mid-1800s, serious efforts were initiated to develop the state-of-the-art for the artificial propagation of several species. Most of the early research on sturgeon aquaculture was unsuccessful. By the early 1900s most culture research programs in the United States were discontinued. In the mid-1920s the Canadians attempted artificial propagation of sturgeon but their results like their predecessors were mostly unsuccessful.

Fishery management projects were initiated on several North American species, primarily for commercial and sport fishing. During this time few attempts were made to advance the progress on developing the state-of-the-art for the artificial propagation of sturgeon. There were specific problems with the various species relative to fish culture and fish management. Some of these problems have been investigated and the research results are included in the following papers.

The symposium on the Biology and Management of Sturgeon of North America was the first of its kind. It brought together scientists from numerous states. They each told their story and hence concluded that most populations of sturgeon in the US and Canada have declined severely in the past 100 years and less is known about their biology and management than was previously thought. Within the papers presented an undertone of one important discipline was constantly referred to – that is the artificial propagation of sturgeon. The development of techniques to successfully culture sturgeon could be applied to stocking programs, gene pool preservation, aquaculture and expanding the biological data base of these important fishes.

This national meeting set a registered attendance record of over 815 people. Undoubtedly the sturgeon symposium contributed to the attraction of the overall program. The one and one-half day session topic, which was unique, attracted an audience of more than 150 people each day. The scientists participating in this symposium represented a geographical range from the west coast to the Great Lakes, to the east coast, to the Gulf of Mexico. Of the twenty-two papers presented seventeen were selected for publication. All papers in this issue were subjected to the normal editorial review process. Our selection of papers includes genetics, reproductive physiology, taxonomy, culture, behavior, development, ecology, and management of six species of sturgeon, which included the white, *Acipenser transmontanus;* lake, *Acipenser fulvescens;* Atlantic, *Acipenser oxyrhynchus;* shovelnose, *Scaphirhynchus platorynchus;* pallid, *Scaphirhynchus albus;* and shortnose, *Acipenser brevirostrum.*

In 1979 the first seeds of interest on sturgeon biology and culture were planted. At this time three states – Wisconsin, California and South Carolina – unbeknownst to one another initiated investigations on developing the state-of-the-art for the artificial propagation of sturgeon. In 1980 the first results of this research were presented at the World Mariculture Society Meetings in New Orleans, Louisiana. At this same meeting an informal gathering of 6 to 8 people was held in a hotel room which comprised the nucleus of sturgeon researchers and research ideas for the future. In the summer of 1980 the first sturgeon research workshop was held at the National Fisheries Center in Leetown, West Virginia, sponsored by the U.S. Fish and Wildlife Service. A few months earlier only a handful of scientists gathered to discuss research and management ideas. At the Leetown meeting more than 40 agency, academic and private industry scientists gathered to exchange information and ideas on sturgeon biology and management.

Shortly afterwards the involvement expanded to include international interests by Norway, Italy, France, China and Israel. U.S. federal funding sources including the University of Wisconsin Sea Grant Institute, California Sea Grant College Program, U.S. Fish and Wildlife Service and National Marine Fisheries Service contributed partial support for various research projects. Along with the expanded geographic interest evolved cooperative research programs between several academic institutions: University of California, University of Wisconsin, University of Washington, University of California, University of Wisconsin, Louisiana State University, and Coker College South Carolina, University of California – and state agencies: Wisconsin, California, Washington, Michigan, Missouri, Oregon and Idaho. In retrospect, little did we realize that a small group of scientists gathered in a hotel room in New Orleans were setting the stage for the 1983 Symposium on the Biology and Management of Sturgeon in North America.

Even though the USSR did not participate in this symposium, we acknowledge the major contribution of Soviet scientists to sturgeon culture and management.

We would like to thank all the authors who contributed to this publication for their cooperation and patience during the review and revision process. We would also like to thank the following individuals who reviewed one or more of the manuscripts: Nancy Auer, Ken Beer, Harold Bergman, Howard Bern, Ernest Brannon, Christopher Calvert, William Craig Clarke, John Conner, Larry Crowder, Mike Dadswell, Edward Donaldson, Lauren Donaldson, Daniel Faber, Neal Foster, Frederick Goetz, Robert Gresswell, John Halver, David Jude, E. David Lane, William Latta, Robert McCauley, Spencer Melecha, Peter Moyle, Nick Parker, Gordon Priegel, James Rice, Robert Stickney, Prudence Talbot, Bruce Taubert, Fred Utter, Victor Vacquier, George Weisel and Steven Yeo. In addition we wish to express our thanks and appreciation to Ratko J. Ristic for creating the logo and drawings which appear on blank pages in this volume.

We also thank the University of Wisconsin-Milwaukee for its support and facilities to host this symposium and a special thanks to the American Fisheries Society for its support and sponsorship of this symposium as part of the program of the 113th Annual Meeting of the American Fisheries Society. Finally, we thank the University of Wisconsin-Milwaukee Center for Great Lakes Studies and the University of California-Davis Department of Animal Science for their support and generosity for allowing us to spend countless hours preparing this publication.

Frederick P. Binkowski
Sergei I. Doroshov

Introduction

The symposium on the 'Biology and Management of Sturgeon' included 22 papers which covered an amazingly broad spectrum of subjects. Six of the seven major sturgeon species found in North America were covered and a number of European and Asian species were also mentioned. The subjects included reproduction, physiology, taxonomy, genetics, growth, life history, population dynamics, culture and management. Papers were presented by fishery scientists from the coasts of the Atlantic, Pacific and Gulf of Mexico as well as the Great Lakes and the Missouri River basin.

Considering that some species of sturgeon have produced only 20 generations since the Europeans first colonized North America it is remarkable that we know as much as we do about this ancient animal. With a life span of the same order of magnitude as that of the elephants, great whales, and even man himself, the difficulties of studying the life history and population dynamics of sturgeon are enormous. Even more challenging is the task of devising and implementing prudent and rational management policies for this fascinating group of fish. Several decades must pass before the results of management practices can be reliably evaluated. Subtle habitat changes which affect only the reproductive cycle and juvenile survival, but not the adults, are difficult to detect for many years. Therefore management practices for sturgeon must lean heavily on the side of conservation and habitat preservation. To establish and maintain such a protectionist policy is usually politically and economically difficult. Nevertheless the more we know about these fish the greater is our chance of achieving and maintaining sound management practices. By highlighting the state of our knowledge on sturgeon this first symposium has made a significant contribution toward this end.

It is heartening to learn that life history studies and strong management programs are underway on both the Atlantic and Pacific coasts as well as in the Great Lakes. In the Columbia River, severe depletion of the white sturgeon by overfishing has been reversed and stocks are on the rebound. At the same time dams have reduced or eliminated recruitment of this and other species in parts of some rivers. One species, the pallid sturgeon, has been classified as rare in Missouri. A more serious situation exists for the shortnose sturgeon, which has been classified as an endangered species on the Atlantic coast. The shovelnose, a species which spends most of its life in rivers, has a fairly stable population in the northcentral part of the U.S. Under careful management Wisconsin's lake sturgeon fishery in Lake Winnebago over the past 30 years has increased in both harvest and total stock. In the case of the sea run Atlantic sturgeon little is known about their oceanic migration and homing behavior. Consequently it is difficult to design a comprehensive management plan.

The demand for sturgeon is increasing. In 1979 the commercial harvest was 240 tons and the sport catch was about 210 tons. On the Columbia River white sturgeon has become the primary target of sportfishermen and in Wisconsin the Lake Winnebago lake sturgeon is sought during the regulated winter spearfishing season. The commercial market in the U.S. for sturgeon as a gourmet food fish is quite strong and growing. In addition there are good opportunities to expand this market in France, Italy, Israel, China and Japan. Since sturgeon are anadromous fish the potential for ranching operations also exists. This approach will probably require the augmentation of the sea run stock by stocking of young fish. Young sturgeon are also in great demand as aquarium fish. All of these demands point to a need for the development of efficient hatchery and culture practices. As pointed out by Doroshov and others the raising of sturgeon in captivity is limited by a lack of knowledge of their behavior, reproduction, genetics, physiology, nutrition and pathology.

Sturgeon rearing has really just begun in the U.S. There are presently culture programs in at least five states – California, Wisconsin, Michigan, Missouri and South Carolina. In the Soviet Union successful sturgeon culture programs have been underway for more than 25 years. In the U.S. husbandry practices are just now being worked out, but interest in commercial production is high. There is little doubt that sturgeon aquaculture will eventually become a commercial reality.

Sturgeon research by its very nature must be long-term. Sound management and particularly habitat preservation is crucially dependent on research into reproduction, early life history, physiology, and population dynamics. At present, research programs on sturgeon are spotty. Only a few states maintain an active research program. Nevertheless interest in these species is growing as government management agencies and university groups recognize the value of these fish and the stresses that river habitat modifications have placed upon them. Increased support of sturgeon research will probably have to come largely from government agencies responsible for the management and well being of the resource. Advances in culture techniques and commercial feasibility will no doubt come from support by the Sea Grant Program and private industry.

Fred P. Binkowski and Sergei I. Doroshov are to be complimented on organizing this first symposium on sturgeon. The response by scientists from all parts of the U.S. and Canada made the meeting a great success. Not only were they able to exchange their results and experiences but also they developed a proposal to form a consortium of state agencies and institutions to formulate research plans for the protection and rehabilitation of the white sturgeon in the Columbia River basin. Given the baseline and impetus of this first symposium, we can now expect future meetings with the attendant research stimulation and management insights so desperately needed for these valuable but little known fishes.

Successful though it was, this first sturgeon symposium would have been greatly strengthened by the participation of European and Asian scientists. There are advanced and long-standing sturgeon research and culture programs underway in U.S.S.R. Let us hope that the next sturgeon meeting can truly be an international one.

Robert A. Ragotzkie
Sea Grant Institute,
University of Wisconsin,
Madison, WI 53705, U.S.A.

Gamete interaction in the white sturgeon *Acipenser transmontanus*: a morphological and physiological review

Gary N. Cherr & Wallis H. Clark, Jr.
Department of Obstetrics and Gynecology and Bodega Marine Laboratory, University of California, Davis, CA 95616, U.S.A.
Address correspondence to: Gary N. Cherr, Department of Obstetrics and Gynecology, Division of Reproductive Biology and Medicine, School of Medicine, University of California, Davis, CA 95616, U.S.A.

Keywords: Sperm, Egg, Fertilization, Acrosome reaction, Micropyles, Chorion, Jelly, Protease, Glycoproteins, Egg water

Synopsis

Sturgeon gametes differ from those of most fish in that the sperm possess acrosomes that undergo exocytosis and filament formation while the eggs possess numerous micropyles. *Acipenser transmontanus* eggs are encased by multilayered envelopes that consist of outer adhesive jelly coats and three structured layers interior to the jelly. The glycoprotein jelly layer only becomes adhesive upon exposure to freshwater. The layer interior to the jelly, layer 3, is the other carbohydrate-containing component of the egg envelope. This layer consists of a water-insoluble glycoprotein that, upon freshwater exposure, is hydrolyzed by a trypsin-like protease to yield a water-soluble, lower molecular weight carbohydrate-containing component. This component can be identified in the surrounding medium when unfertilized eggs are incubated in freshwater. This egg water component elicits acrosome reactions only in homologous sperm. The *A. transmontanus* sperm acrosome reaction is a Ca^{++} and/or Mg^{++} dependent event that includes the formation of a 10 μm long fertilization filament. *A. transmontanus* fertilization can occur at low sperm per egg ratios; however, cross-fertilization of *A. transmontanus* eggs with lake sturgeon, *A. fluvescens*, sperm results in a very low number of fertilized eggs, even at high sperm per egg ratios. The morphological, physiological, and biochemical phenomenon reviewed in this paper are related to the environment in which they occur. Also, the possible role of the acrosome and the presence of numerous micropyles are discussed.

Introduction

The cellular events of fertilization have been extensively studied in numerous phyla throughout the animal kingdom. More often than not, these studies have ignored the reproductive strategies of the organism providing the gametes; instead, the investigators have used the cells as models to study events such as membrane fusion, cell motility, and cell-cell recognition (see Epel 1978, Lopo & Vacquier 1981, and Shapiro & Eddy 1980, for reviews). In addition to the above, gametes (particularly sperm) have been carefully examined and categorized providing clues to phylogenetic schemes (Bacetti & Afzelius 1976, Bacetti 1979).

Most cellular research has been conducted using echinoderms and rodents. Echinoderms exhibit external fertilization with the gametes being shed into seawater while mammals exhibit internal fertilization, the gametes being exposed to relatively constant environments. Animals display a broad spectrum of gamete types that can be related to the environment in which they interact. This variation corresponds to the reproductive strategy of the particular animal. For example, invertebrates and lower vertebrates may exhibit either internal or external fertilization and they may be broadcast spawners or brooders.

Within the lower vertebrates the greatest diversity in reproductive strategy and gamete structure occurs in fish. Marine species exhibit both brooding and broadcast spawning while freshwater species are typically brooders. Fish eggs are usually encased by an impenetrable extracellular coat (chorion) that is perforated by a singlefunnel-shaped micropyle at the animal pole which allows sperm direct access to the egg's plasma membrane (see Gilkey 1981). The spermatozoa of most fish are simple spherical cells that lack an acrosome and possess a typical flagellum. Some of the more primitive groups of fishes (Chondrostei, Holostei) exhibit gametes and fertilization patterns that differ from the above. An example is acipenserids (Chondrostei) which possess eggs that have numerous micropyles at the animal pole and have sperm that are elongate with conspicuous acrosomes (see Ginzburg 1968 for a review, Cherr & Clark 1982, 1984b).

This review of gamete interaction in the white sturgeon, *Acipenser transmontanus* (an anadromous fish from the west coast of North America) discusses the functional morphology, physiology, and biochemistry of the temporal and spatial events that occur when sperm meets egg. These events will then be related to the environment in which they occur.

The egg and its investments

Animal eggs are usually encased by extracellular coat(s) that the sperm must traverse prior to gamete fusion. These coats serve a variety of functions that include sperm attraction and swarming (Collins 1976), initiation of the acrosomal reaction (SeGall & Lennarz 1979), egg adhesiveness (Cherr & Clark 1984a), and finally, protection of the developing embryo.

Most fish eggs are covered by a tough proteinaceous envelope or chorion. This investment is impenetrable to sperm and blocks polyspermy since only one sperm can gain access to the oolemma via the single micropyle. White sturgeon eggs are also encased by an impenetrable, multilayered egg envelope; however, this envelope is perforated by numerous micropyles (3–15) at the animal pole (Cherr & Clark 1982).

The mature ovum of *A. transmontanus* is 3.5–4.0 mm in diameter and is heavily pigmented. The egg envelope is a 50 μm thick structure that consists of four morphologically distinct regions (Fig. 1) (Cherr & Clark 1982). The outermost layer (layer 4) is an electron dense jelly coat which extends into the ducts of layer 3 (L3) (Fig. 2a, b). Upon exposure to freshwater, this L4 is released from L3 and hydrates to form a highly adhesive layer (Fig. 1). When this jelly is isolated and analyzed by sodium dodecyl sulfate polyacrylamide gel electrophoresis (SDS-PAGE), a single protein band of 110 000 daltons (KD) is apparent. This protein is glycosylated as demonstrated by carbohydrate staining of gels (Cherr & Clark 1984a). The jelly layer is 88.1% protein and 11.9% carbohydrate. The carbohydrate portion of the jelly contains sialic acid as demonstrated by staining of the jelly with the fluorescently tagged lectin *Limulus polyhemus* agglutinin (LPA) and by the identification of sialic acid in isolated jelly (Cherr & Clark 1984a).

The adhesive jelly coat is present throughout early development and is probably important in anchoring the embryo to a specific region of river bottom. The physiological mechanisms of jelly release have been examined (Cherr & Clark 1984a). When sturgeon eggs are immersed in freshwater (river water, distilled water, or freshwater with a defined composition) the amorphous jelly layer is released from the surface and the ducts of L3 and hydrates within 5 min (Fig. 2c, d). This phenomenon does not occur in coelomic fluid or physiological salines (unpublished data). However, incubation of eggs for 30 min in freshwater that lacks Ca^{++} and Mg^{++} ions will also inhibit jelly hydration. These experiments differ from incubations in distilled water since the eggs are extensively washed with $Ca^{++} - Mg^{++}$ free freshwater containing the chelating agent ethylene diamine tetraacetic acid (EDTA). This experiment indicates that Ca^{++} and/or Mg^{++} is probably required for jelly release and hydration.

It has also been found that by incubating eggs for

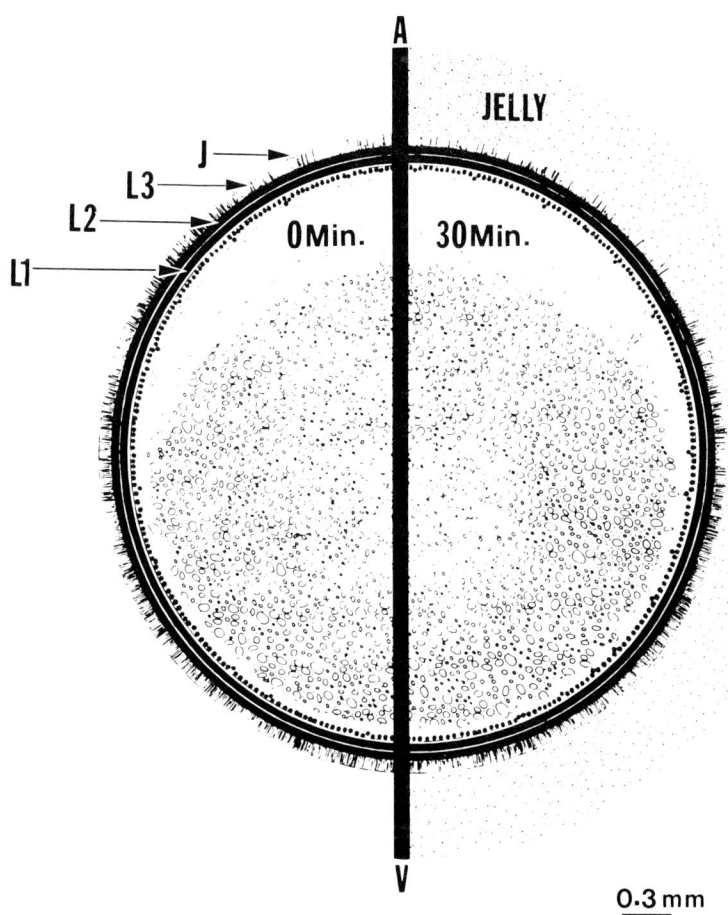

Fig. 1. Schematic of the *A. transmontanus* egg before (0 min) and after (30 min) freshwater exposure. Note the four layers of the egg envelope (J, L3, L2, L1) and the hydrated jelly following immersion in freshwater. A = animal pole; V = vegetal pole.

30 min in freshwater containing either general protease inhibitors or specific inhibitors of trypsin-like enzymes, jelly release could be inhibited or blocked. Inhibitors of chymotrypsin-like enzymes did not inhibit egg adhesiveness. Urea extracts of eggs incubated in the protease inhibitors that block jelly release do not contain the 110KD glycoprotein that can typically be isolated from eggs that are adhesive. Thus, the 110KD glycoprotein probably originates from a higher molecular weight complex that is acted on by a trypsin-like enzyme. It is believed that this protease resides in the egg envelope since its presence it not dependent on fertilization or egg activation (Cherr & Clark 1984a).

Sturgeon egg jelly differs from the jelly of other animal eggs. In contrast to amphibian eggs, the jelly of *A. transmontanus* eggs is non-oviducal in origin and is not required for fertilization. In addition, egg adhesiveness is independent of fertilization and egg activation, unlike some invertebrate eggs (Fallon & Austin 1967). Finally, the vegetal pole of *A. transmontanus* eggs appears more adhesive than the animal pole (unpublished data). This differential adhesiveness may assure attachment of the vegetal pole to a substrate, leaving the animal pole oriented upward. This could be important for orientation of the micropyles prior to fertilization as well as the establishment of the dorso-ventral axis during embryonic development.

L3, the investment coat interior to the jelly coat also contains carbohydrate. This layer stains positively with periodic acid-Schiff (PAS) and dansyl hydrazine (Cherr & Clark 1985). Intense labeling of L3 occurs when the fluorescently tagged

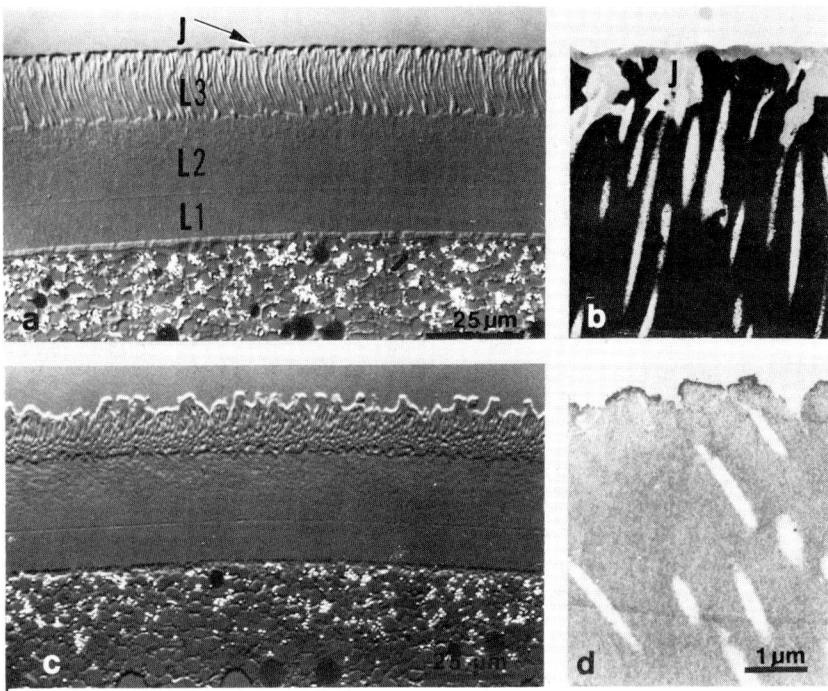

Fig. 2. a – Light micrograph of the egg envelope showing the four layers (L1, L2, L3, jelly), J = jelly; *b* – transmission electron micrograph of L3 and the jelly (J) that extends into the ducts. *c* – Light micrograph of the egg envelope following a 30 min exposure to freshwater (note that the thin jelly layer is gone and the ducts of L3 have expanded); *d* – transmission electron micrograph of L3 following freshwater exposure (note the absence of jelly and the loss of electron density in L3).

lectin, *Ricinus communis* agglutinin (RCA) is incubated with *A. transmontanus* egg envelopes, indicating that D-galactose is a major component. L3 of the *A. transmontanus* egg envelope is comparable to the jelly layer described in eggs of other sturgeon (see Ginzburg 1968). However, L3 does not hydrate or dissolve to form the white sturgeon egg jelly (Fig. 3A). In addition, urea removal of *A. transmontanus* egg jelly does not morphologically alter L3 (unpublished data).

L3 is firmly anchored to its adjacent layer, L2, via acellular 'bioscrews' (Fig. 3a, b) (Cherr & Clark 1982). These unusual helical structures are composed of thin filaments from L2. The integrity of L3 and the screws of L2 remain following fertilization, cleavage, and early development. It has been observed that ovulated eggs of sturgeon collected from salt water prior to the spawning migration have premature separation of L3 and jelly from L2 – in these eggs, the screws are incomplete and appear as stubby projections. The only obvious morphological change occurring in L3 following a 30 min exposure to freshwater is a loss of electron density (Fig. 2d). The screws from L2 are also indirectly responsible for retention of the adhesive jelly coat since this layer resides external to L3.

Layers 1 and 2 are composed of fibrils that possess filamentous substructural elements (Fig. 3a). Neither of these layers stain positively for carbohydrate and exhibit no change through fertilization and cleavage. These layers are comparable to chorions of other fish eggs due to their structure and tough protective properties.

Acipenserid eggs differ from other fish eggs in that they possess numerous micropyles located within a 200 μm diameter area at the animal pole (Fig. 3c) (Cherr & Clark 1982). The number of micropyles in *A. transmontanus* eggs ranges from 3–15 with an average of 7. Each micropyle is generally funnel-shaped, tapering to the diameter of a single sperm (Cherr & Clark 1982) (Fig. 3d). These micropyles provide the sperm direct access to the

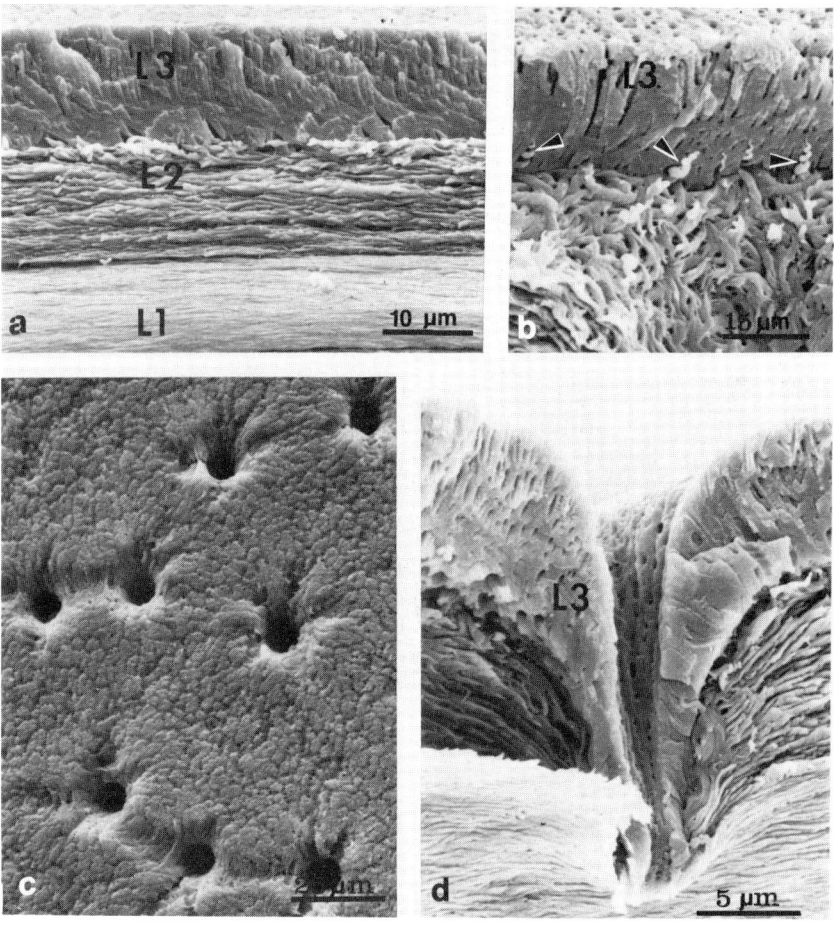

Fig. 3. a – Scanning electron micrograph (SEM) of the *A. transmontanus* egg envelope; *b* – SEM demonstrating the 'bio-screws' (arrows) of L2 that anchor L3 and the jelly to the egg; *c* – SEM of the egg surface and the micropyles at the animal pole; *d* – SEM of a fracture through a micropyle down to L1.

egg's oolemma. The length of the micropylar canal is 50 μm; the outer 30–40 μm is lined by L3 (Fig. 3d). Thus, the terminal portion of the micropylar canal is through L1. The jelly layer only partially extends into the micropylar canal. The inner opening of each micropyle (at L1) is closely apposed to the egg surface (oolemma).

Due to the presence of numerous micropyles, it was believed that sturgeon eggs were normally penetrated by more than one sperm (polyspermy) (see Ginzburg 1968 for a review). The result of polyspermy in most systems is developmental arrest. It is now believed that polyspermy in sturgeons is abnormal. This topic will be discussed later.

The spermatozoon

In general, the apical organelle of sperm, the acrosome, is considered to be responsible for enabling sperm to traverse investment coats surrounding an egg. Most sperm undergo an acrosome reaction in response to an egg component; this acrosomal reaction exposes the contents of the acrosome, which includes enzymes and binding proteins (see Dan 1967, Lopo 1983). Sea urchin sperm typically undergo an acrosome reaction upon contacting the egg's jelly coat (see Shapiro & Eddy 1980 for a review). This acrosome reaction includes exocytosis of the acrosomal vesicle and the formation of a fertilization filament or acrosomal process.

The enzymes released during the acrosome reaction are presumed necessary for penetration of egg investments. Many functions have been proposed for the acrosomal process; however, little direct evidence for any of these functions have been documented.

While many vertebrate sperm possess acrosomes, very few exhibit an acrosomal process. Among vertebrates, only lampreys and sturgeons are reported to possess sperm that form fertilization filaments during the acrosome reaction (Jaana & Yamamoto 1981, Detlaf & Ginzburg 1963, Cherr & Clark 1984b).

The presence and/or structural diversity in the acrosome is believed to be related to the barriers through which sperm must penetrate in order to fuse with an oolemma. Eggs with impenetrable coats are provided with one or more micropyles that allow sperm direct access to the egg's membrane. Many freshwater spawning species typically possess sperm that lack an acrosome. For example, teleost fish are believed to have evolved in freshwater and later invaded the marine environment. The absence of an acrosome thus, in teleost fish sperm, probably corresponds to both the presence of micropyles in the eggs and the fishes' origin from the freshwater environment (Baccetti & Afzelius 1976).

It is believed that the low ionic strength of freshwater creates a hyposmotic condition that tends to disrupt membranes of sperm (Morisawa & Suzuki 1980). Since the membranes of the acrosome are relatively unstable, the acrosomal reaction could occur in a temporally and spatially uncontrolled manner (Baccetti & Afzelius 1976). This is apparently true in lamprey sperm; unless the sperm contacts the egg, a spontaneus acrosomal reaction will occur at the termination of motility (Jaana & Yamamoto 1981). Sturgeon sperm, however, do not display this type of acrosomal instability since only a low percentage (6%) of sperm that have ceased motility after freshwater dilution undergo the acrosomal reaction (Cherr & Clark 1984b).

Lamprey eggs are covered by a penetrable envelope that lacks micropyles; thus, the acrosome of lamprey sperm is presumably responsible for sperm penetration (Kille 1960). Sturgeon eggs, however, possess an impenetrable envelope which is perforated by numerous micropyles that provide sperm direct access to the oolemma. Therefore, the presence of acrosomes and long (10 μm) acrosomal processes in sturgeon spermatozoa is puzzling and does not agree with models of acrosomal evolution (Baccetti & Afzelius 1976, Baccetti 1979).

A. transmontanus sperm (Fig. 4) become motile upon dilution in freshwater; these sperm are not motile in the coelomic fluid of the female. The burst of motility that occurs when the sperm are diluted in water is immediate but decreases by 3 min. By 5 min post-dilution, most sperm have ceased all directed motility. The duration of sturgeon sperm motility in freshwater and the lack of motility in female coelomic fluid is a dramatic departure from the behavior of salmonid sperm. These cells exhibit extended motility in female coelomic fluid and exhibit membrane disruption within 90 s following freshwater dilution (Morisawa & Suzuki 1980), while sturgeon sperm remain structurally intact for up to 10 min in freshwater (unpublished data).

As mentioned previously, sturgeon sperm are unique in that the acrosomal reaction (Fig. 5a, b) appears to be a temporarily controlled event, even though it occurs in freshwater. We have examined the ionic requirements of the acrosomal reaction in *A. transmontanus* sperm and have found that the ionic requirements are similar to those of marine invertebrate sperm (Cherr & Clark 1984b) (Table 1). This is surprising since freshwater is lower in ionic strength than seawater. Also, ion concentrations can vary greatly when compared to the relatively stable marine environment. Apparently, these cells have retained mechanisms for activation that one would have expected to change in sperm

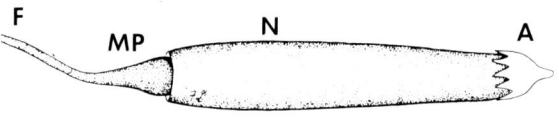

Fig. 4. Schematic of the *A. transmontanus* spermatozoon. A = acrosome; N = nucleus; MP = midpiece; F = flagellum.

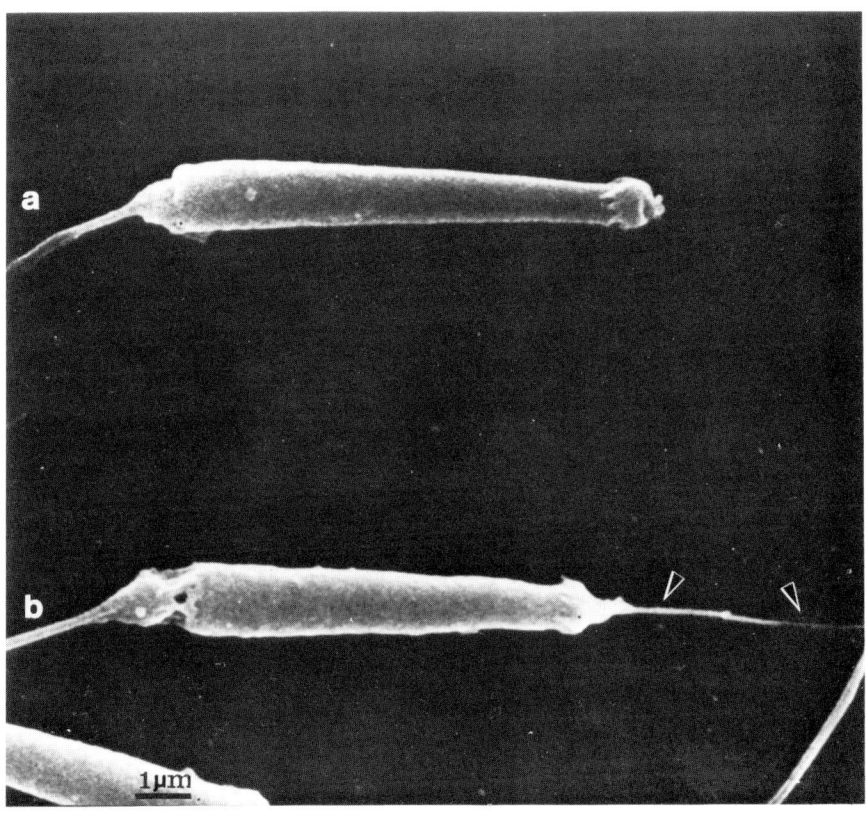

Fig. 5. a – SEM of sperm prior to undergoing the acrosome reaction; *b* – *A. transmontanus* sperm following induction of the acrosome reaction. Note the acrosomal process (arrows).

Table 1. Percentage of acrosome reactions generated in *A. transmontanus* sperm by acrosome reaction inducers. Percentages are expressed as means (n = 3) ± S.D. The concentration of A23187 was 200 μg ml^{-1}.

Treatment	% Acrosome reaction
Freshwater pH 7.6	6 ± 2.30
Freshwater + 1% DMSO	4 ± 1.03
Freshwater + A23187	74 ± 13.22
Ca^{++} – Mg^{++} free freshwater + A23187	6 ± 4.36
Ca^{++} – free freshwater + A23187	19 ± 1.73
Mg^{++} – free freshwater + A23187	19 ± 6.99
High Ca^{++} – freshwater (30 mM)	48 ± 2.90
High Mg^{++} – freshwater (30 mM)	6 ± 1.99
Freshwater pH 5.5 + A23187	9 ± 3.98
Freshwater pH 9.5	65 ± 13.32
Egg water	47 ± 2.51

from a hypertonic medium (seawater) to sperm from a hypotonic medium (freshwater).

As reported in other sturgeon species, homologous egg water (water which has contained eggs) induces the acrosomal reaction in white sturgeon sperm (Cherr & Clark 1984b, 1985). Since an understanding of the possible role(s) of the acrosome in the fertilization process was of interest, components responsible for inducing the sturgeon sperm acrosomal reaction were analyzed and localized. This topic is discussed below (Cherr & Clark 1985).

Sperm-egg interaction

Unlike most freshwater spawning fishes, sturgeon are broadcast spawners and release their eggs over large areas of river bottom. The eggs become adhesive and probably remain within the spawning area. However, sturgeon eggs can be exposed to freshwater for hours and still remain fertilizable (Ginzburg 1968, Cherr & Clark unpublished data). This differs from nesting fishes such as salmonids whose eggs activate and cannot be fertilized following a brief exposure to freshwater (Kussa 1950). As mentioned previously, sturgeon sperm differ from salmonid sperm in that the duration of motility is longer and the sperm are elongate and possess acrosomes. The benefit of increased duration of motility is of obvious importance to a broadcast spawner, however, the presence of an acrosome in sturgeon sperm and its function during gamete interaction is less obvious. Thus the component from the egg that induces the acrosome reaction as well as its origin were investigated in order to understand acrosomal activation in sturgeon sperm.

The component in egg water which is released from the *A. transmontanus* egg envelope (following immersion in freshwater) that induces the acrosomal reaction is a 66 KD glycoprotein (Cherr & Clark 1985). Electrophoretic analysis of the *A. transmontanus* egg envelope prior to freshwater exposure demonstrated that the major component is a 70 KD glycoprotein; this is the precursor of the 66 KD inducer in egg water. This is the only carbohydrate containing component (other than jelly) in the egg envelope and thus corresponds to L3. In order to determine both acrosome reaction inducing activity and specificity, the 70 KD glycoprotein was isolated from the gel and incubated with *A. transmontanus* and lake sturgeon, *A. fluvescens,* sperm. This component induced acrosomal reactions in only *A. transmontanus* sperm. Egg water is also species specific in its ability to elicit an acrosomal reaction since it has no effect on lake sturgeon sperm. The 66 KD dalton glycoprotein was only observed in envelope preparations from eggs exposed to freshwater. Sperm were also incubated with isolated jelly and it was found that jelly possesses no acrosome reaction inducing activity (Cherr & Clark 1985).

By using enzyme inhibitors, it was established that the appearance of the 66 KD inducer in the 30 min egg envelope and surrounding medium is due to the activity of a trypsin-like protease. This is probably the same enzyme that is responsible for jelly release. The 70 KD glycoprotein is in a water-insoluble form and remains in the egg envelope; however, both the 70 KD and 66 KD glycoproteins are species specific in their inducing activity (Cherr & Clark 1985).

Since the natural induction of the sturgeon sperm acrosome reaction appeared to be species-specific, *A. transmontanus* eggs were fertilized with *A. fluvescens* sperm in order to determine if cross-fertilization could occur. At concentrations normally used for in vitro fertilization of homologous sturgeon eggs (1:200 sperm dilution; 10^3 sperm per egg), no fertilization occurred. Only at concentrations of 10^6 sperm per egg did a low percentage (5%) of fertilization occur. Furthermore, homologous in vitro fertilization results in numerous sperm in and around the egg's micropyles (Fig. 6a, b). However, the heterologous fertilization resulted in a complete absence of sperm on the egg (Fig. 6c). We believe that the small percentage of hybrid fertilization that occurred when extremely high concentrations of sperm were used was due to the fact that a low percentage (6%) of sperm undergo the acrosome reaction upon dilution in freshwater; these may be the successful sperm.

Experiments to determine the minimal *A. transmontanus* sperm concentration necessary for successful in vitro fertilization of *A. transmontanus* eggs were conducted in the laboratory using petri dishes (unpublished data). Fifty eggs were placed in each dish, washed 3 times, and 20 ml of freshwater was added. To these dishes, concentrated sperm was added such that the final concentration of sperm in the various dishes ranged from 2.5×10^6 ml^{-1} (10^6 sperm per egg) to 25 ml^{-1} (10 sperm per egg). While a low percentage of eggs from the highest concentration appeared polyspermic, fertilization was greater than 50% at all concentrations used. Even at the lowest concentration of sperm used (10 sperm per egg), fertilization was 53%.

From these experiments, one may hypothesize possible functions of numerous micropyles in sturgeon eggs. As discussed previously, sturgeon are broadcast spawners; therefore, sperm concen-

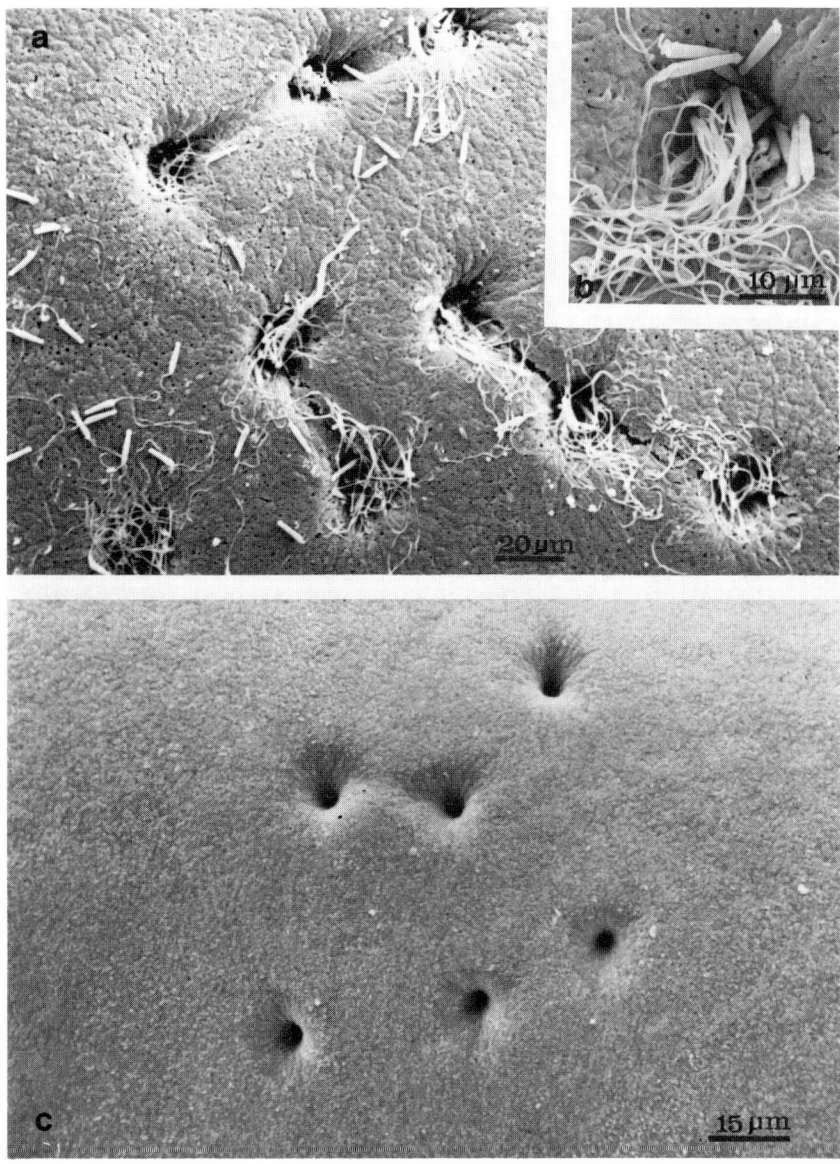

Fig. 6. a – SEM of the surface of an *A. transmontanus* egg 1 minute after insemination with *A. transmontanus* sperm (10^5 sperm per egg); *b* – higher magnification of *A. transmontanus* sperm in a micropyle; *c* – SEM of *A. transmontanus* egg following insemination with *A. fulvescens* sperm (10^6 sperm per egg).

trations would be expected to be lower in the surrounding medium than in fish that are nesters. As such, numerous micropyles would ensure successful fertilization even with low concentrations of sperm. Another possible explanation for the presence of numerous micropyles could be the necessity for the sturgeon acrosome reaction to occur in the micropyles. Since sturgeon eggs are spawned into fast moving waters, the soluble inducer of the acrosome reaction would probably be too dilute to affect sperm when they are in the surrounding medium. In addition, sturgeon spermatozoa must traverse a 50 μm thick egg envelope via the micropyles in order to contact the oolemma. Thus, the generation of a 10 μm long acrosomal process outside of a micropyle would not enable this fragile structure to contact the oolemma. It is assumed that the acrosome reaction must normally occur within the L3 portion of the micropylar canals. It is within this region that concentrations of the 66 KD inducer would be greatest; furthermore, the sperm are approximately 10 μm away from the egg's membrane at this point. This is the same length as the sperm's fertilization filament. A requirement for the acrosome reaction to occur in the micropylar canal could explain the presence of numerous micropyles since acrosome reacted sperm entering the micropyles would not be capable of fertilizing the egg and might eliminate potentially successful, unreacted sperm from having access to the micropylar canals and the oolemma. Thus, numerous micropyles could increase chances of successful fertilization (Cherr & Clark 1985).

As mentioned previously, the presence of numerous micropyles in sturgeon eggs does not appear to increase the chances of polyspermy. In *A. transmontanus*, extremely high concentrations of sperm result in a low percentage (<10%) of eggs

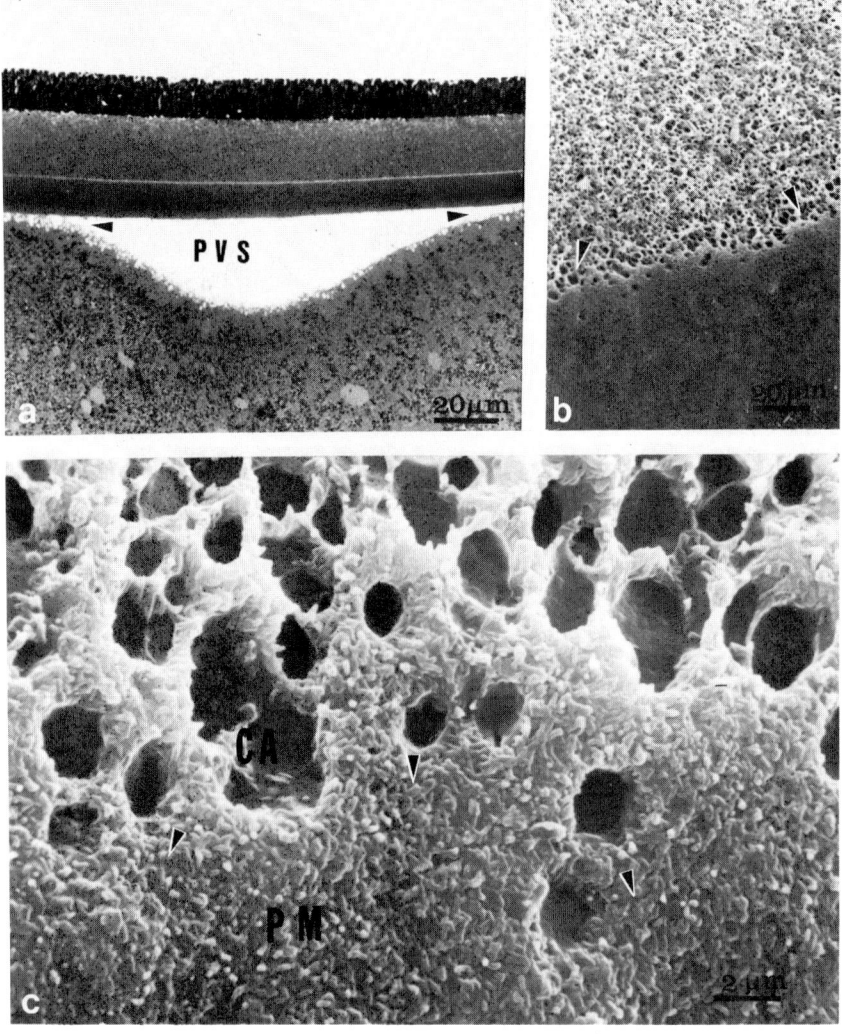

Fig. 7. a – Light micrograph of animal pole region 20 seconds post-insemination. Note the cortical reaction (arrows) and the rapidly forming perivitelline space (PVS); *b* – SEM of the egg's plasma membrane 20 seconds post-insemination. The cortical reaction has occurred at the upper portion of the micrograph and is proceeding (arrows) towards the vegetal pole; *c* – higher magnification of the cortical reaction. Note the vesicles or cortical alveoli (CA) that have fused with the plasma membrane (PM). The wave of exocytosis is proceeding towards the vegetal pole (lower portion of micrograph) (arrows).

that are polyspermic and do not develop (unpublished data). However, polyspermy is rarely observed at normal sperm concentrations. The block to polyspermy in sturgeon is apparently the extremely fast cortical reaction that is triggered when the sperm contacts the oolemma (Fig. 7a). This reaction involves the exocytosis of numerous cortical granules or alveoli that lie beneath the oolemma (Fig. 7b, c) such that the egg's plasma membrane is altered, the cortical granule material rapidly sealing the inner micropylar openings. This cortical reaction appears quick enough to block supernumerary sperm from penetrating the oolemma (unpublished data).

Conclusions

This discussion of sturgeon gametes and their interaction has emphasized morphological and physiological phenomena associated with a freshwater, broadcast-spawning animal. These phenomena differ from most of the freshwater spawning fishes (salmonids, cyprinids) in the following ways: (1) *A. transmontanus* eggs possess numerous micropyles; (2) can remain in freshwater for hours without activation; (3) releases a soluble factor that induces the acrosome reaction in homologous sperm; (4) *A. transmontanus* sperm exhibit prolonged motility in freshwater; (5) are elongated with conspicuous acrosomes; and (6) undergo an acrosomal reaction that includes the formation of a fertilization filament. When one looks at the environment in which sturgeon gametes interact, much of the above is logical. However, the presence of an acrosome and an acrosomal process in the sperm and numerous micropyles in the egg would seem to be contradictory.

The acrosome reaction in sturgeon does appear to be important as an interspecific block to fertilization. While most brooders and nesters would not be susceptible to this problem due to behavioral adaptations (pairing, nesting, etc.), a broadcast spawner such as *A. transmontanus* could be susceptible to interspecific fertilization. Thus, the requirement for the acrosome reaction as a prerequisite for fertilization and its species specific induction would seem to be an explanation for this phenomenon.

Gametes from acipenserids would appear to provide excellent material for understanding gamete function and evolution. While reproductive strategies at the gamete level differ among taxa, this variety of patterns can probably be correlated with the various spawning conditions and behavior of the animal. As such, sturgeon demonstrate reproductive adaptations (at the gamete level) for broadcast spawning in the river environment.

Acknowledgements

The authors wish to express thanks to F.J. Griffin for his fine illustrations. This research was supported by USDC 81-ABH-001101, NMFS Contract 80-ABC00017, and NOAA, National Sea Grant College Program, Department of Commerce #NA80AA-D-00120, through the California Sea Grant Program and California State Resources Agency, Project #RA45.

References cited

Baccetti, B. & B.A. Afzelius. 1976. The biology of the sperm cell. Monographs in Developmental Biology, Vol. 10. 254 pp.

Baccetti, B. 1979. The evolution of the acrosomal complex. pp 305–329. *In:* D.W. Fawcett & J.M. Bedford (ed.) The Spermatozoon, Urban and Schwarzenberg, Baltimore.

Cherr, G.N. & W.H. Clark, Jr. 1982. Fine structure of the envelope and micropyles in the eggs of the white sturgeon, *Acipenser transmontanus* Richardson. Develop. Growth and Differ. 24: 341–352.

Cherr, G.N. & W.H. Clark, Jr. 1984a. Jelly release in the eggs of the white sturgeon, *Acipenser transmontanus*: an enzymatically mediated event. J. Exp. Zool. 230: 145–149.

Cherr, G.N. & W.H. Clark, Jr. 1984b. An acrosome reaction in sperm from the white sturgeon, *Acipenser transmontanus*. J. Exp. Zool. 232: 129–139.

Cherr, G.N. & W.H. Clark, Jr. 1985. An egg envelope component induces the acrosome reaction in sturgeon sperm. J. Exp. Zool. (in press).

Collins, F. 1976. A reevaluation of the fertilizin hypothesis of sperm agglutination and the description of a novel form of sperm adhesion. Develop. Biol. 49: 381–394.

Dan, J.C. 1967. Acrosome reaction and lysins. pp. 237–288. *In:* C.B. Metz & Monroy (ed.) Fertilization, Comparative Morphology, Biochemistry, and Immunology, Volume 1, Academic Press, New York.

Detlaf, T.A. & A.S. Ginzburg. 1954. Embryonic development of sturgeon in connection with artificial propagation. Trudy Inst. of Animal Morphology Acad. of Sci. USSR 47, Moscow. 204 pp.

Detlaf, T.A. & A.S. Ginzburg. 1963. Acrosome reaction in sturgeons and the role of calcium ions in the union of gametes. Doklady Acad. Nauck SSSR 153: 1461–1464.

Epel, D. 1978. Mechanisms of activation of sperm and egg during fertilization of sea urchin gametes. Curr. Topics Dev. Biol. 12: 185–246.

Fallon, J.F. & C.R. Austin. 1967. Fine structure of gametes of *Nereis limbata* before and after interaction. J. Exp. Zool. 166: 225–242.

Gilkey, J.C. 1981. Mechanisms of fertilization in fishes. Amer. Zool. 21: 359–375.

Ginzburg, A.S. 1968. Fertilization in fishes and the problem of polyspermy. Nauka Press. Moskva. 175 pag. Israel program for scientific translations. 1972.

Jaana, H. & T.S. Yamamoto. 1981. The ultrastructure of spermatozoa with a note on the formation of the acrosomal filament in the lamprey, *Lampetra japonica*. Jap. J. Ichtyol. 28: 135–147.

Kille, R.A. 1960. Fertilization of the lamprey egg. Exptl. Cell. Res. 20: 12–27.

Kusa, M. 1950. Physiological analysis of fertilization in the egg of the salmon *Oncorynchus keta*. I-Why are the eggs not fertilized in isotonic ringer solution? Ann. Zool.- Japon. 2: 2–28.

Lopo, A.C. 1983. Sperm-egg interactions in invertebrates. pp. 269–324. *In*: J.F. Hartman (ed.) Mechanisms and Control Fertilization, Academic Press, New York.

Lopo, A.C. & V.D. Vacquier. 1981. Gamete interaction during sea urchin fertilization: a model for studying the molecular details of animal fertilization. pp. 199–232. *In*: L. Mastroianni & J.D. Biggers (ed.) Fertilization and Early Development, Plenum Press, New York.

Morisawa, M & K. Suzuki, 1980. Osmolatity and potassium ions: their roles in initiation of sperm motility in teleosts. Science 210: 11454–1146.

SeGall, G.K. & W.J. Lennarz. 1979. Chemical characterization of the component of the jelly coat from sea urchin eggs responsible for induction of the acrosome reaction. Dev. Biol. 71: 33–48.

Shapiro, B.M. & E.M. Eddy. 1980. When sperm meets egg: biochemical mechanisms of gamete interaction. Int. Rev. Cytol. 66: 257–302.

Received 15.5.1984   Accepted 5.1.1985

Osmoregulation in juvenile and adult white sturgeon, *Acipenser transmontanus*

Maryann McEnroe & Joseph J. Cech, Jr.
Departments of Zoology and Wildlife and Fisheries Biology, University of California, Davis, CA 95616, U.S.A.

Keywords: Chondrostean, Plasma electrolytes, Salinity, Fish

Synopsis

Blood samples from cannulated young adult (2.5–15 kg) white sturgeon, acclimated to San Francisco Bay water (24 ppt) had plasma values of 248.8 ± 13.5 mOsm kg^{-1} H_2O, $[Na^+] = 125 \pm 8.0$ mEq l^{-1}, $[K^+] = 2.6 \pm 0.8$ mEq l^{-1} and $[CL^-] = 122 \pm 3.0$ mEq l^{-1}. Freshwater acclimated sturgeon had an osmolality of 236 ± 7, $[Na^+] = 131.6 + 4.4$, $[K^+] = 2.5 \pm 0.7$ and $[CL^-] = 110.6 \pm 3.6$. Freshwater acclimated fish gradually exposed to sea water (increase of 5 ppt h^{-1}) had higher plasma osmolalities than did the bay water acclimated fish. These young adult sturgeon are able to tolerate transfer from fresh water to sea water as well as gradual transfer from sea water to fresh water. Plasma electrolytes in transferred fish are regulated, but tend to differ from long term acclimated fish at the same salinities. There is a gradual increase in the upper salinity tolerance (abrupt transfer) of juvenile white sturgeon with weight: 5–10 ppt for 0.4–0.9 g fish, 10–15 ppt for 0.7–1.8 g fish, and 15 ppt for 4.9–50.0 g fish. The ability of juveniles to regulate plasma osmolality is limited. The young adult fish are able to tolerate higher salinities (35 ppt) than juvenile sturgeon but probably are also characterized by low activity of the necessary ion exchange mechanisms in the gills which permit rapid adjustment of blood electrolytes with graduate change in external salinity.

Introduction

Acipenser transmontanus is a chondrostean, a group of ancient actinopterygian fish which evolved at least 200 million years ago. *A. transmontanus* is found along the Pacific coast of North America from British Columbia to California and is anadromous, migrating up the larger rivers along the coast (the Fraser, Columbia, Umpqua, and Sacramento Rivers) to spawn. It is believed that these fish reach sexual maturity between fifteen and twenty years of age, and may spawn every two to ten years thereafter, depending on age and physiological state (Doroshov 1985). In the fall these sturgeon enter the estuaries associated with the rivers and slowly migrate upstream over the next several months. Spawning occurs during the winter months (February-March). Sturgeon juveniles remain in fresh water from several months to several years, depending on the species (Doroshov 1985). However, it is not known how long *A. transmontanus* juveniles remain in a freshwater environment, or when they start their seaward migration. Thus, the natural history of *A. transmontanus* suggests several interesting physiological problems: How do they osmoregulate? When can the juveniles successfully tolerate sea water?

Although there have been many investigations of osmoregulation in adult teleosts, there have been fewer on the more 'primitive' fish groups.

This is especially true of the Chondrostei, as a result of their large size, and the difficulty of obtaining them. Three previous studies on sturgeon osmoregulation (Magnin 1962, Urist & Van de Putte 1967, Potts & Rudy 1972) have indicated that sturgeon are hypo/hyperosmotic regulators, as are the teleosts. However, a paucity of data and a lack of documentation regarding environmental conditions prompt further studies into chondrostean osmoregulation. This is especially true concerning the development of osmoregulatory capabilities in young fish. Virtually all of the studies in this area have been focused on the anadromous salmonids, as a result of their extensive hatchery culture and economic importance.

In order to determine how adult white sturgeon osmoregulate, and how this capacity for osmoregulation changes with development, our study was composed of two parts: a study of plasma electrolytes of adult specimens, and a study of salinity tolerance in juveniles.

Materials and methods

Studies on adult sturgeon

Twenty adult *Acipenser transmontanus* were snagged in San Francisco Bay at the initiation of the spawning migration during the winters of 1981–1982 and 1982–83. The fish were of both sexes and early stages of gonadal development as determined after experimentation. After capture, the fish were transported to the Tiburon Laboratory of the National Marine Fisheries Service, and placed in tanks with flow-through, aerated, bay water having a salinity of 22–26 parts per thousand (ppt).

During the winter of 1981–1982 all experiments were conducted at the Tiburon Laboratory. Ten sturgeon were held in fiberglass tanks (2.14 m × 0.91 m × 0.50 m depth) with a constant flow of San Francisco Bay water (12–14°C) having a salinity range of 22–26 ppt. The water temperature (YSI Model 51B thermistor), salinity (AO Goldberg refractometer), and dissolved oxygen (YSI Model 51B electrode and meter) were monitored frequently. The fish were allowed to recover from the stress of capture for one week, and then were cannulated as described below. These fish were used to obtain the bay water acclimated electrolyte values.

The following winter (1982–1983) ten sturgeon (*A. transmontanus*) were caught as previously described and slowly acclimated to fresh water over several days. They were then transported from Tiburon, California, to the Institute of Ecology at the University of California, Davis (UCD) in fresh water. In Davis, the fish were allowed to recover from the stress of transport, and to further acclimate to fresh water, for another two weeks. They were then cannulated as described below and placed in the fiberglass tanks with a flow-through supply of non-chlorinated fresh water (11–14°C) and allowed to recover for 2–3 days before the initiation of the experiment. The temperature, salinity, and dissolved oxygen of the water were monitored daily. Upon initiation of the experiment initial freshwater acclimated blood samples were taken (as described below) and the plasma frozen for later analysis. The salinity was then increased from fresh water (0 ppt) to sea water (33–35 ppt) over a two-day period (increases of 5 ppt h^{-1}), by starting a flow of dissolved and aerated Instant Ocean brine into the tank (11–14°C). On the first day of the experiment the salinity was increased to 22 ppt over eight h and left at that salinity for 24 h. Blood samples were taken at each end of this 24 h period at 22 ppt. The next day the flow of the brine solution was again initiated, and the salinity increased to 33–35 ppt over several hours. The fish were kept at 33–35 ppt for three days in a static system. Blood samples were taken when the salinity reached 33–35 ppt and then taken every 24 h for the next 3 d. The water was gradually replaced with fresh 'sea water' on the second day. At the end of the 3 d period the salinity was decreased to 0 ppt (5 ppt h^{-1}) and blood samples taken. Blood samples were then taken after 24 h and one week in fresh water.

Cannulation technique

The fish were allowed to recover from the stress of capture and/or transport for at least several days prior to ventral aortic or ventricular cannulation. The fish to be cannulated was placed ventral surface up into a fabric stretcher having a hood at one end for the sturgeon's head. A hose carrying water for gill ventilation was placed into the fish's mouth. In this position the fish would usually remain still. No anesthetic was used. An eighteen gauge, thin-wall 5 cm hypodermic (cannulation) needle was inserted dorsally at the isthmus, into the ventral aorta or ventricle. A slight negative pressure on the attached syringe pulled deoxygenated blood into the syringe, and a pulsatile flow was observed. At this point the syringe was removed, and approximately 5 cm of the 1.3 m length of PE 50 cannula tubing (0.580 mm I.D. × 0.965 mm O.D.) was inserted through the cannulation needle. The cannulation needle was then removed over the cannula tubing which was filled with a heparinized saline solution (ammonium heparin in 0.9% saline). The blood in the cannula was flushed back into the fish with the heparinized saline. The cannula was secured to the ventral surface and side of the fish with sutures so that the distal portion of the cannula floated in the water. After ensuring that blood could be drawn through the cannula, it was refilled with heparinized saline and tied off, and the fish returned to its tank.

The cannulation technique proved effective for many of the experimental specimens. Many completed the increase/decrease in salinity cycle successfully and were subsequently released into San Francisco Bay or the Sacramento River. However, in several specimens the cannula pulled out during the experiment, limiting the data gathered on these fish.

Blood sampling technique

Since the purpose of the cannulation was to obtain repeated blood samples without stressing the fish, care was taken not to disturb or excite the fish when blood samples were taken. Blood samples were taken by cutting the knot on the cannula, removing the saline, and attaching a pre-heparinized (dry ammonium heparin crystals) syringe and 23 gauge needle to the cannula. Blood samples of 2 ml volume were taken during the winter of 1981–1982,

and 1 ml samples during the winter of 1982–1983.

After the samples were obtained, microhematocrits were determined and the rest of the blood was centrifuged at 4500 rpm for 5 min. Most of the plasma was removed and frozen. The remaining plasma and packed red blood cells were remixed and injected into the fish, after the saline had been removed from the cannula. Thus, repeated blood samples could be obtained without significantly altering either the oxygen carrying capacity or the buffering capacity of the blood.

Electrolyte determinations

The frozen plasma samples were later thawed and then immediately analyzed for plasma sodium, potassium, chloride, and osmolality. Plasma sodium and potassium were analyzed using a IL 143 flame photometer, chloride with Radiometer CMT-10 chloride titrator, and osmolality with a Wescor 5100B vapor pressure osmometer.

Studies on juvenile sturgeon

Juvenile white sturgeon were obtained from several spawns of wild caught sturgeon at the aquaculture facility at the Institute of Ecology, UCD. The fish were transported to our laboratory in the Department of Wildlife and Fisheries Biology, UCD. They were maintained in fiberglass tanks having a continuous flow of aerated, stripped, and non-chlorinated well water until use. The fish were fed the diet on which they had been raised (either tubifex worms or Oregon Moist Pellets) supplemented with live brine shrimp.

Upon initiation of the experiment, the fish were weighed in a tared beaker with water. Five weight groups were used: 0.4–0.9 g, 0.7–1.5 g, 0.9–1.8 g, 4.9–9.5 g, and 18.0–56.0 g. The fish were abruptly transferred to an insulated, plastic cooler which contained water of the test salinity at the same temperature as the fresh water tank (17–19°C). Water for the experiments was made up using non-chlorinated well water and Instant Ocean artificial sea salts. The salinity was measured with the refractometer. The water was aerated and mixed overnight before initiation of the experiment.

Ten fish of each size group were tested at each salinity (0 ppt, 5 ppt, 10 ppt, 15 ppt, 25 ppt, 30 ppt, and 35 ppt), and survival was monitored for three days. 'Survival' at the test salinity was determined as survival for three days or longer. This duration was chosen as preliminary experiments indicated that those juvenile sturgeon which survived three days at the test salinity would survive indefinitely at that salinity. In tests where fish survived 24 h, test water was daily replaced with water of the test salinity after feeding. The experiments on the smallest fish (0.4–0.9 g and 0.7–1.5 g) were conducted in March, several weeks after spawning. The 0.9–9.5 g fish and the 18.0–56.0 g fish were run in November, and the following September, respectively.

To test whether pre-acclimation to 10 ppt or 15 ppt for one week would increase survival at higher salinities, 70 fish of the 4.9–9.5 range were abruptly transferred to the lower salinities (40 fish to 10 ppt and 30 fish to 15 ppt), and maintained for 1 week with daily water changes. At the end of the week-long acclimation period, 10 fish from each group were transferred to water of the same salinity (either 10 ppt or 15 ppt) as well as to higher salinities: 15 ppt, 25 ppt, and 35 ppt. Survival was monitored for three days.

Plasma samples were obtained from fish of the largest size class used (18.0–56.0 g) after 9 h in either 15 ppt or 25 ppt by sacrificing the fish and obtaining blood by caudal section. The osmolalities of the samples were determined with the vapor pressure osmometer.

Results

The bay water acclimated sturgeon were hypoosmotic to bay water, and the freshwater acclimated sturgeon were hyperosmotic to fresh water (Table 1). The plasma osmolality and electrolyte values of the bay water and freshwater acclimated fish were similar (Table 1). No significant difference was found in the plasma concentrations of Na^+ and K^+ and the osmolality when comparing bay water acclimated and freshwater acclimated adult sturgeon. There was a significant decrease ($P<0.001$, t-test) in the concentration of Cl^- in freshwater acclimated adults compared with bay water acclimated fish (Table 1). When the freshwater acclimated sturgeon were gradually exposed to increased salinity up to sea water over a two day period, and then held in sea water for 3 days there was a continual rise in all the measured plasma variables (Table 2). An analysis of variance showed that there were no significant differences between the freshwater and 22 ppt (24 h) concentrations of Na^+, Cl^-, or the osmality in these fish. However, there were significant differences ($P<0.05$) between the freshwater and both the initial and 3

Table 1. Plasma electrolyte values ($\bar{X} \pm SD$) of bay water and freshwater acclimated Acipenser transmontanus.

	Salinity (Time at that salinity)		Level of significance
	22–26 ppt (>2 wk) (bay water acclimated)	0 ppt (> several wk) (freshwater acclimated)	
Na^+ (mEq l^{-1})	125.0 ± 8.0	131.6 ± 4.4	N.S.
K^+ (mEq l^{-1})	2.6 ± 0.8	2.5 ± 0.7	N.S.
Cl^- (mEq l^{-1})	122.0 ± 3.0	110.6 ± 3.6	$P<0.001$
Osmo (mOsm kg^{-1})	248.8 ± 13.5	236.1 ± 7.1	N.S.
Number of fish	7	7	
Osmo (mOsm kg^{-1}) of water	approx. 700	<100	

Sample sizes are smaller than indicated in text due to the unfortunate proclivity of these fish to either jump out of their tank and/or knock out standpipes. In either case, the fish died.

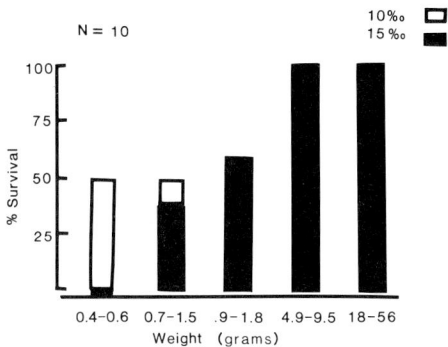

Fig. 1. Percent survival of several size classes of juvenile white sturgeon with abrupt transfer to either 10 ppt (open bars) or 15 ppt (closed bars) salinity.

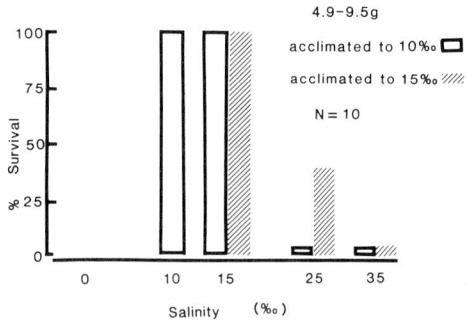

Fig. 2. Percent survival of 4.9–9.5 g white sturgeon at various salinities with one week acclimation to either 10 ppt (open bars) or 15 ppt (hatched bars) salinity.

day sea water values of Na^+, Cl^-, and osmolality. The K^+ concentrations of the freshwater acclimated fish compared with the 3 day sea water values were significantly different ($P<0.05$).

There was a gradual increase in the survival of juvenile *A. transmontanus* after abrupt transfer to 10 ppt, 15 ppt, with increased size (weight) as shown in Figure 1. All fish of all size groups survived transfer to 0 ppt and 5 ppt. Some fish below 1.8 g could not withstand abrupt transfer to 10 ppt whereas some fish below 4.9 g did not survive transfer to 15 ppt. In addition, no juvenile fish survived direct transfer to salinities of 25 ppt and above. Blood samples obtained from 18–56 g fish after nine h in 15 ppt showed a small rise in plasma osmolality from a fresh water mean of 245 ± 1 to 306 ± 6. In contrast, the plasma osmolality of the sacrificed (but dying) fish at 25 ppt had risen dramatically to 410 ± 19. Pre-acclimation to 10 ppt for one week did not increase survival at 25 ppt, or 35 ppt. However, pre-acclimation to 15 ppt for one week did slightly increase survival at 25 ppt, but not 35 ppt (Fig. 2).

Discussion

The white sturgeon, *Acipenser transmontanus*, is one of twenty-nine extant species of Chondrostei (Nelson 1976). The chondrosteans are the oldest and most primitive of the actinopterygians, and are believed to have evolved at least two hundred million years ago (Nelson 1976), and possibly as long as four hundred million years ago (Løvtrup 1977) from the palaeoscinoids. Doroshov (1985) has divided many of the living sturgeon species into three categories: anadromous, semi-anadromous, and landlocked (Table 3). *A. transmontanus* is a semi-

Table 2. Plasma electrolyte values ($\bar{X} \pm SD$) of freshwater acclimated *Acipenser transmontanus* in fresh water and during salinity increase to sea water.

	Salinity (Time at that salinity)			
	0 ppt (> several wk)	22 ppt (24 h)	33–35 ppt (<1 h)	33–35 ppt (3 d)
Na^+ (mEq l^{-1})	131.6 ± 4.4	145.8 ± 4.4	167.0 ± 18.1	200.8 ± 12.6
K^+ (mEq l^{-1})	2.5 ± 0.7	2.4 ± 0.3	2.8 ± 0.5	3.9 ± 0.5
Cl^- (mEq l^{-1})	110.6 ± 3.6	119.4 ± 3.0	138.9 ± 14.0	169.3 ± 11.6
Osmo (mOsm l^{-1})	236.1 ± 7.1	272.5 ± 8.9	327.0 ± 27.7	385.0 ± 18.9*
Number of fish*	7	4	4	4

anadromous fish, which spends most of its adult life in sea water, but close to shore, and migrates into fresh water to spawn.

Previous studies on sturgeon osmoregulation (Magnin 1962, Urist & Van de Putte 1967, Potts & Rudy 1972) have indicated that sturgeon are hypo/hyperosmotic regulators, as are the euryhaline and diadromous teleosts. Magnin (1962) captured species of sturgeon (*Acipenser oxyrhynchus*, *A. sturio*, and *A. fulvescens*), took blood samples, and analyzed the plasma electrolytes. Urist & Van de Putte carried out a similar study on *A. transmontanus* caught in Suisun Marsh and San Francisco Bay. Potts & Rudy (1972) measured ^{22}Na$^+$ efflux in one *A. transmontanus*, and several *A. medirostris* in sea water, fresh water, and during the transition. Urist & Van de Putte (1967) reported that *A. transmontanus* caught in San Francisco Bay had similar plasma electrolyte and osmolality values to the *A. transmontanus* caught in Suisun Marsh. Our data are in partial agreement with their findings. However, we found a significant difference in the chloride levels between our freshwater and bay water fish, whereas they did not. We also found that both bay water and freshwater acclimated sturgeon had lower osmolalities (248.8 ± 13.5, 236.1 ± 7.1, respectively) than did their single data points for bay water and Suisun Marsh caught fish (275 mOsm). These discrepancies between data sets may be due to the fact that our freshwater values came from fish held in fresh water, and not Suisun Marsh, which has salinity variations from 0 ppt to 12 ppt, depending on the season and the yearly rainfall (personal observation). Unfortunately, Urist & Van de Putte (1967) did not report the salinities at which their fish were caught for either the Bay or Marsh. Studies are underway to determine plasma electrolyte values for sea water acclimated sturgeon.

The dramatic rise in plasma electrolytes of the freshwater acclimated fish gradually exposed to sea water is not surprising as this has also been found to be true of diadromous American eels which are not endocrinologically prepared for the transfer from fresh water to sea water (Ball et al. 1971, Forrest et al. 1973a, b). These eels require about ten days after the transfer from fresh water to sea water for the plasma electrolytes to decrease to those of sea water acclimated eels. This is a result of the freshwater fish not having the ion excreting capacity necessary to maintain blood electrolyte values at sea water acclimated values. In eels, the hormone cortisol is responsible for many of the changes which accompany sea water acclimation (Forrest et al. 1973a, b, Hirano 1980) while in salmonids both cortisol and thyroxine are involved in sea water acclimation (Loretz et al. 1982). Sturgeon, like eels and salmonids, are diadromous and undergo mi-

Table 3. Classification of some of the Acipenseridae by spawning habitat. Classification scheme and data from Doroshov (1985).

	Classification	Habitat	Spawning location	Species
I.	Truly anadromous species	oceans	migrate up rivers to spawn	1. *Acipenser oxyrhynchus* 2. *A. sturio* 3. *A. medirostris*
II.	Anadromous or semi-anadromous species	associated with estuarine bays or impoundments species 4, 5, 6, can enter sea water in coastal areas; species 6, 7, 8 can form land-locked populations	migrate up rivers to spawn	4. *Huso huso* 5. *Acipenser transmontanus* 6. *A. guldenstaedti* 7. *A. nudiventris* 8. *A. fulvescens*
III.	Completely landlocked populations	inhabit large fresh water lakes and reservoirs (species 10, 11); or live in riverine and lake environments (species 12 and 13)	migrate upstream to spawn	9. *A. fulvescens* 10. *A. ruthenus* 11. *A. baeri* 12. *Pseudoscaphirhynchus* sp. 13. *Polyodon spathula*

grations from fresh water to sea water at certain times of the year. Because this is a seasonal, and therefore predictable event, it would not be surprising if there is also a hormonal component to sea water acclimation in sturgeon.

Cortisol has been implicated in the ability of both euryhaline (Doneen 1976, Doneen & Berg 1974, Foskett et al. 1981, Foskett et al. 1983) and anadromous (Mayer et al. 1967, Forrest et al. 1979a, b) teleosts to adapt to increased salinities. Cortisol appears to act by stimulating morphological changes in the gills of both euryhaline (Foskett et al. 1981) and anadromous (Doyle & Epstein 1972) fish, as well as in the gut of anadromous fish (Hirano et al. 1975, Hirano & Mayer-Gostan 1976, Hirano 1980). We are currently pursing studies on the role of cortisol in seawater adaptation in adult and juvenile white sturgeon.

In our study of the development of salinity tolerance in juvenile *A. transmontanus*, we found increased salinity tolerance with size. However, we did not find a critical body size above which salinity tolerance is independent of body size as has been reported for striped mullet (Nordlie et al. 1982). Both starry flounder, *Platichthys stellatus*, and striped mullet, *Mugil cephalus*, showed increased osmoregulatory ability with increased size (Hickman 1959, Nordlie et al. 1982). The salinity tolerance of white sturgeon between 56 g and 15 kg body weight has not been investigated. The minor difference between the freshwater osmolality data in the juvenile sturgeon (245 ± 1) and the adult sturgeon (236 ± 7) in the present study may have resulted from the two different blood sampling techniques used.

Several studies have found body size to be important in the development of salinity tolerance in juvenile salmonids (Parry 1958, 1960, 1961, Conte & Wagner 1965, Wagner et al. 1969, Farmer et al. 1978). Parry (1958, 1960) concluded that body size, and not age, was the crucial factor in the development of salinity tolerance in salmonids. On the other hand Clarke (1982) believes that it is a maturational event rather than body size that is important. He found that the body size at which the five species of salmonids he studied (chum, sockeye, chinook and coho salmon, and steelhead trout) exhibited 'optimal hypoosmoregulatory capacity' varying from 0.6 g for chum salmon to 35.0 g for steelhead.

If the development of salinity tolerance is dependent on a maturational event, rather than on surface to volume ratio (i.e., body size) then it would be of interest to determine the nature of the maturational events. Since chloride cells are necessary for survival in a hyperosmotic environment (Foskett et al. 1981, Foskett & Scheffey 1982), it may be that the development of chloride cells in the gills is one of a series of important maturational events. Presently we are studying the development of chloride cells in juvenile sturgeon and the role of cortisol in the development of this cell type in the gills of juvenile sturgeon.

Acknowledgements

We thank Abe and Angelo Cuanang, Eddie Tavasieff, George Monaco, and the captain and crew of the R/V Stephanie K II for assistance in adult specimen collection. We thank Serge Doroshov, Ken Beer, and Yuan Wang for their cooperation in providing juvenile specimens. We also thank Norm Abramson, Jeannette Whipple, and Brian Jarvis for providing us with facilities and technical support at the NOAA, National Marine Fisheries Service, Tiburon Laboratory. Support for the work came in part from the University of California, Davis, Aquaculture Program, Departments of Wildlife and Fisheries Biology and Zoology, the U.C. Agricultural Experiment Station, Institute of Ecology, and NOAA, National Sea Grant College Program, Department of Commerce, under grant number NA80AA-D-00120 through the California Sea Grant College Program, and in part by the California State Resources Agency, project number R/F-90. The U.S. Government is authorized to reproduce and distribute for governmental purposes. The manuscript was typed by Ellen Tani and Donna Raymond.

References cited

Ball, J.J., I. Chester Jones, M.E. Forester, G. Hargreaves, E.F. Hawkins & K.P. Milne. 1971. Measurement of plasma cortisol levels in the eel, *Anguilla anguilla* in relation to osmotic adjustments. J. Endocrinol. 50: 75–96.

Clarke, W.C. 1982. Evaluation of the seawater challenge test as an index of marine survival. Aquaculture 28: 177–183.

Conte, F.P. & H.H. Wagner. 1965. Development of osmotic and ionic regulation in juvenile steelhead trout, *Salmo gairdneri*. Comp. Biochem. Physiol. 14: 603–620.

Doneen, B.A. 1976. Water movement in the urinary bladder of the gobiid teleost *Gillichthys mirabilis* in response to prolactin and cortisol. Gen. Comp. Endocrinol. 28: 33–41.

Doneen, B.A. & H.A. Bern. 1974. *In vitro* effect of prolactin and cortisol on water permeability of the urinary bladder of the teleost *Gillichthys mirabilis*. J. Exp. Zool. 187: 173–179.

Doroshov, S. 1985. The biology and culture of sturgeon. *In*: J. Muir & R. Roberts (ed.) Recent Advances in Aquaculture, Croon Helm Publ., London. (in press).

Doyle, W.L. & F.H. Epstein. 1972. Effect of cortisol treatment and osmotic adaptation on the chloride cell in the eel, *Anguilla rostrata*. Cytobiologie 6: 58–73.

Farmer, G.J., J.A. Rilter & D. Ashfield. 1978. Seawater adaptation and parr-smolt transformation of juvenile atlantic salmon, *Salmo salar*. J. Fish. Res. Board Can. 35: 93–10.

Forrest, J.N., A.D. Cohen, D.A. Schon & F.H. Epstein. 1973a. Na^+ transport and Na-K-ATPase in the gills during adaptation to seawater: effects of cortisol. Amer. J. Physiol. 224: 709–713.

Forrest, J.N., W.C. Mackay, B. Gallager & F.H. Epstein. 1973b. Plasma cortisol response to saltwater adaptation in the American eel, *Anguilla rostrata*. Amer. J. Physiol. 224: 714–717.

Foskett, J.K., C.D. Logsdon, T. Turner, T.E. Machen & H.A. Bern. 1981. Differentiation of the choride cell extrusion mechanism during seawater adaptation of a teleost fish, the cichlid *Sarothodon mossambicus*. J. Exp. Biol. 94: 209–224.

Foskett, J.K. & C. Scheffey. 1982. The chloride cell: definitive identification as the salt secretory cell in teleosts. Science 215: 164–165.

Foskett, J.K., H.A. Bern, T.E. Machen & M. Conner. 1983. Chloride cells and the hormonal control of teleost fish osmoregulation. J. Exp. Biol. 106: 255–281.

Hickmann, C.P., Jr. 1959. The osmoregulatory role of the thyroid gland in the starry flounder, *Platichthyes stellatus*. Can. J. Zool. 37: 997–1060.

Hirano, T. & N. Mayer-Gostan. 1976. The eel esophagus as an osmoregulatory organ. Proc. Nat. Acad. Sci. USA 73: 1348–1350.

Hirano, T., M. Morisawa, M.T. Ando & S. Utida. 1975. Adaptive changes in ion and water transport mechanisms in the eel intestine. pp. 301–317. *In*: J.W.L. Robinson (ed.) Intestinal Ion Transport, M.T.P., London.

Hirano, T. 1980. Effects of cortisol and prolactin in ion permeability of the eel esophagus. pp. 143–150. *In*: B. Lahlou (ed.) Epithelial Transport in the Lower Vertebrates, Cambridge University Press, Cambridge.

Loretz, C.A., N.L. Collie, N.H. Richman III & H.A. Bern. 1982. Osmoregulatory changes accompanying smoltification in coho salmon. Aquaculture 28: 67–74.

Løvtrup, S. 1977. The phylogeny of the vertebrates. John Wiley and Sons, New York. 330 pp.

Magnin, E. 1962. Rescherches sur la systematique et la biologie des Acipenserides. Res. Piscoles, Ministere de l'Agriculture, Paris, Annls. Stat. Cent. d'Hydrobiol. Appl. 9: 7–244.

Mayer, N., J. Maetz, D.K.O. Chan, M. Forster & I. Chester Jones. 1967. Cortisol, a sodium excreting factor in the eel (*Anguilla anguilla* L.) adapted to sea water. Nature 214: 1118–1120.

Nelson, J.S. 1976. Fishes of the world. John Wiley & Sons, New York. 416 pp.

Nordlie, F.G., W.A. Szelistowski & W.C. Nordlie. 1982. Ontogenesis of osmotic regulation in the striped mullet, *Mugil cephalus* (L.). J. Fish Biol. 20: 79–86.

Oide, M. & S. Utida. 1967. Changes in water and ion transport in isolated intestine of the eel during salt-water adaptation and migration. Mar. Biol 1: 102–106.

Parry, G. 1958. Size and osmoregulation in salmonid fishes. Nature 161: 1218–1219.

Parry, G. 1960. The development of salinity tolerance in the salmon, *Salmo salar* (L.), and some related species. J. Exp. Biol. 37: 411–427.

Parry, G. 1961. Osmotic and ionic changes in the muscle of migrating salmonids. J. Exp. Biol. 38: 411–427.

Potts, W.T.W & P.P. Rudy. 1972. Aspects of osmotic and ionic regulation in the sturgeon. J. Exp. Biol. 56: 703–715.

Urist, M.E. & K.A. Van de Putte. 1967. Comparative biochemistry of the blood of fishes. pp. 271–285. *In*: P.W. Gilbert, R.W. Mathenson & D.P. Hall (ed.) Sharks, Skates, and Rays, Johns Hopkins Press, Baltimore.

Wagner, H.H., F.P. Conte & J.L. Fessler. 1969. Development of osmotic and ionic regulation in two races of chinook salmon, *Oncorhynchus tshawytscha*. Comp. Biochem. Physiol. 29: 325–341.

Received 15.5.1984   Accepted 26.2.1985

Digestive and feeding characteristics of the chondrosteans

Randal K. Buddington[1] & Jay P. Christofferson[2]
[1] *Department of Physiology, UCLA Medical Center, Los Angeles, CA 90024, U.S.A.*
[2] *Department of Biology, California State University, Stanislaus Turlock, CA 95381, U.S.A.*

Keywords: Sturgeons, Digestive system, Anatomy, Physiology, Acipenserids, Polyodontids

Synopsis

Structure and function of the digestive system and feeding characteristics of the chondrosteans are reviewed. Although the group exhibits a wide diversity of feeding habits including piscivory, benthophagy, and planktivory, they are principally carnivores throughout their life history. Examination of digestive system structure reveals the basic structure to be similar among the species with some modification to accomodate the different food types. For the species studied, composition of the adult digestive enzyme complement is consistent with the carnivorous habits with proteases dominating and only low levels of carbohydrase activity. There are three secretory phases during development of the digestive system. Each corresponds with different food habits of the different life-history intervals. Vision is apparently not utilized for feeding in any interval. Instead food is recognized and located primarily by gustation, olfaction, textural qualities, and, possibly, electroreception.

Introduction

Although attempts to rear chondrosteans were initiated at the turn of the century, the first successful large scale production of sturgeon juveniles, developed by Soviet scientists, was not accomplished until approximately 30 years ago (Doroshov, personal communications). Currently Soviet hatcheries are dependent on a labor intensive industry to supply live feeds for sturgeon rearing (Milstein 1972). However, because of inadequate supplies and nutritional deficiencies that can result from their use it is not possible to grow the fish to market size. As a result the fish are released at about 3 g with further growth occurring in the wild. The availability of effective manufactured rations would increase hatchery efficiency and would facilitate growth of sturgeon under intensive aquaculture. A previous lack of information regarding acipenserid digestive physiology, nutritional requirements, and feeding characteristics has hindered the formulation of such rations. However, during the last two decades some of this data has become available.

Chondrosteans are carnivorous throughout their life histories. Larvae are pelagic zooplanktivores (Baranova & Miroshnichenko 1969) whereas food habits of juveniles and adults are more diverse and range from filter feeding (*Polyodon spathula*, Weisel 1973) to piscivory (*Huso huso*). Most adult chondrosteans, though, are opportunistic benthic carnivores with molluscs, crustaceans, insect larvae, other benthic invertebrates and, occasionally, fish comprising the majority of the diet. Although vision does not appear to be utilized for feeding (Sbikin 1974), actual mechanisms of food recognition and location remain unclear. The low growth and survival elicited by manufactured rations in earlier studies (e.g. Charlon & Williot 1978) may actually be related to a poor understanding of the feeding behaviors and characteristics, not nutrient composition, of the diets. Development of effective rations and associated feeding techniques will also require a thorough understanding of the mechanisms and physiology of digestion as well as nutritional requirements.

The present review is devoted to chondrostean and especially acipenserid feeding characteristics and the associated sensory adaptations, digestive system anatomy and physiology, development of manufactured rations, and feeding techniques. Deficiencies in our knowledge are revealed and potential avenues of research are indicated. Speculations are also extended concerning applications of the data toward understanding the natural history of the chondrosteans and the development of culture technology.

Digestive system structure

Digestive system anatomy was described for a number of chondrosteans during the last century (see Hopkins 1895) and recently there has been a renewal of interest (Weisel 1973, 1979, Buddington & Doroshov 1985c). Between these periods most of the efforts have been devoted to descriptions of early development. Although most descriptions of the adult gut are based on only a few specimens, they do reveal general structural similarities. The alimentary canal is short (between 70–100% of fork length) with a single loop which is formed by the stomach. Identifiable regions include a short esophagus, secretory forestomach, non-secretory pylorus, intestine with attached pyloric caecum, spiral valve, and a short rectum.

The esophagus (5% of alimentary canal length) extends from the pharynx to just prior to the entrance of the physostome swim bladder. Internally, the limits of the esophagus coincide with the presence of numerous posteriorly directed conical papillae. The papillae are larger in macrophagous species relative to the microphagous paddlefish (Hopkins 1895) and appear to assist in movement of food particles.

The stomach is composed of two distinct regions which together comprise 40–50% of the gut length (Buddington & Doroshov, 1985c). The anterior forestomach is elongated (30–40% of total gut length) and within macrophagous species is capable of distending to 3–5 times the resting state. In contrast, the forestomach walls of the paddlefish are thicker and less distensible (Hopkins 1895) and also include four parallel rows of fatty rods which may be an adaptation for movement of small food particles (Weisel 1973). Such rods are not present in macrophagous species. Longitudinal rugae, present when the forestomach is undistended, disappear when the stomach is filled with food.

The muscle layers of the pyloric region are hypertrophied forming what is often termed a gizzard, particularly in species which feed on hard food items. As a result, the gizzard is capable of triturating diet components and thereby compensates for a complete lack of dentition (jaw, oral, or pharyngeal) in adults. For example, although intact crustaceans and molluscs are observed in the forestomach, only fragments are recovered from the intestine and spiral valve (personal observations RKB). In contrast, there is comparatively little muscle hypertrophy in the paddlefish. Rugae are lacking in the gizzard and the internal lining has a velvety appearance. The gizzard terminates at the pyloric sphincter.

Although the intestine is straight for most of its length, a curvature is present prior to the connection with the spiral valve. External coloration of the intestine, as well as the rest of the alimentary canal, ranges from almost white to black and may be related to degree of parasitic infection (Buddington & Doroshov 1985c). The internal structure is a network of interconnecting ridges with deep crypts which form a complex topographical relief. The bile duct enters on the tip of a papilla located within a few centimeters of the pyloric sphincter in adult fish (Macallum 1886, Buddington & Doroshov 1985c) with the entrance to the pyloric caecum at approximately the same level. The pyloric caecum is a flattened, triangular structure positioned between the intestine and stomach. Its outer edge is lobiform. This may represent the fusion of numerous caecae into a single organ (Macallum 1886, Weisel 1979). A large number of internal channels within the caecum corroborate this hypothesis. The channels anastomose into a single opening which communicates with the intestine.

The spiral valve represents 50–70% of the postgastric alimentary canal and has a greater diameter than the intestine (Buddington & Doroshov 1985c). Backflow of the chyme into the intestine is prevented by a valve within the s-shaped curvature which separates the intestine from the spiral valve. The complex folding of the mucosa observed in the intestine is present but is not as deep. A central typhosole traverses the length of the spiral valve. It is connected to the wall by a continuous sheet of spiralling tissue thereby forming the characteristic valves. Although individual variation is present, the number of valves is species specific and ranges from 4–5 in *Scaphirhynchus* (Weisel 1979) to 6–8 in most other acipenserids. The valves effectively increase surface area by a factor of 3.3 (Andrew & Hickman 1974). Acipenserids therefore increase surface area of the gut by the presence of a spiral valve, a pyloric caecum, and a complex mucosal structure.

A short rectum (2–3% of gut length) extends from the end of the spiral valve to the anal vent. Longitudinal rugae are present internally.

The liver is composed of three principle lobes. The right and largest lobe is located along the lateral edge of the intestine and also covers a portion of the gizzard. A large bulb-shaped gall bladder is located within a notch on the ventral surface. The left lobe extends across the transverse segment of the stomach whereas the smallest medial lobe is positioned between the stomach and intestine in conjunction with the pyloric caecum.

The pancreas is a diffuse organ disseminated throughout the body cavity and can also be detected within various organs. Small sample sizes in conjunction with individual variation and the incorporation of pancreatic tissues into numerous organs, particularly the liver and spleen, has caused some discrepancies regarding location of the pancreas. This was also noted by Macallum 1886). Pancreatic tissues are also associated with blood vessels in the body cavity, especially those along the intestine and spiral valve. Portions of the pancreas can be observed under the intestinal serosa of some specimens (Macallum 1886, Weisel 1979, Buddington & Doroshov 1985c). Secretions from the diffuse pancreas appear to be collected into a single duct which joins with the bile duct once it enters the intestine and are also released from the tip of the papilla described earlier.

Microscopic anatomy

There are relatively few descriptions of the histological organization of the chondrostean digestive system. The following is based on the observations of Macallum (1886), Hopkins (1895), Weisel (1973, 1979), and Buddington & Doroshov (1985c). Each alimentary canal region exhibits modifications of the basic structure which appear to represent adaptations for specific digestive functions. The gut is composed of three tissue layers: mucosa; muscularis; and serosa or adventitia. There are only a few reports of a muscularis in isolated gut sections (Macallum 1886).

Esophageal tissue structure is consistent with a food transport function. The epithelium is stratified (5–10 cells thick) with a surface layer that can be squamous, cuboidal, or columnar. Goblet cells are scattered throughout whereas taste buds, when present, are located on tips of the papillae in the proximal esophagus. The mucosal connective tissue is thick (50–70% of wall thickness) and dense. In conjunction with intense goblet cell mucous production, it would protect against physical damage during ingestion of hard and large food items. Longitudinally oriented myelinated nerve fibers and fat deposits are located deep in the mucosa. In paddlefish the fat rods are developed into four longitudinal structures. The muscularis consists of a single circular layer of striated muscle. A thick adventitia surrounds the esophagus. In some larger specimens of white sturgeon there is an external layer of columnar cells which contain pigment.

In addition to food storage, chemical digestion is initiated in the forestomach as indicated by the presence of gastric glands. The greatest densities are just distal to the pneumatic duct. Gastric pits lead to the branched, multicellular glands which are composed of a single secretory cell type consistent with those of other lower vertebrates (Smith 1961). Eosinophilic, tryptophan positive particles indicate the presence of pepsinogen granules. There is a simple columnar epithelium dominated by goblet cells lining the necks of the glands and the lumenal mucosa. Apical borders of the epithelial cells possess microvilli. Ciliated cells are also present throughout the alimentary canal, except in the gizzard and rectum, of all life history stages. Although ciliated cells occur in larval guts of other fish, their continued existence in adults is considered a primitive trait (Magid 1975). They may assist in food movements during larval stages. In older fish possessing a well-developed muscularis the cilia may instead enhance distribution of digestive secretions and mucus. There is only a circular layer of smooth muscle in the forestomach whereas the longitudinal layer is also evident by the level of the pneumatic duct. A thin serosa is present with an outer squamous cell layer.

Smooth muscle is the dominant tissue of the gizzard and comprises greater than 80% of the dorsal and ventral walls. The lateral walls are thinner with less muscle hypertrophication. Connective tissue within the mucosa is very dense and is covered by a highly folded simple epithelium composed of almost exclusively goblet cells. Since a specialized keratin lining or other analogous structure is lacking, the connective tissue and copious mucous secretions probably function to protect against mechanical damage during trituration. Microvilli on the goblet cells appear to assist with maintenance of the mucous layer, not nutrient transport. Amino acid and sugar uptakes, measured by an in vitro method, are very low in the stomach (personal observations RKB).

The simple intestinal epithelium can be histologically differentiated into two distinct types, each with apparently different functions. Lumenally exposed portions are composed predominantly of goblet cells which are larger than those of the stomach. Apical borders of the cells have either microvilli or cilia. Within the crypts and deep channels of the mucosa the epithelium consists of columnar cells with only a few goblet cells which suggests absorption is localized there. Movement of chyme down to the absorptive region is enhanced by extensions of the muscularis into the mucosal ridges. Contractions of the smooth muscle would also decrease the unstirred layers and thereby enhance nutrient uptake. Multicellular glands have not been reported. The muscularis consists of both circular and longitudinal smooth muscle layers and is surrounded by a thin serosa which contains pigments in older fish.

Structure of the pyloric caecum is virtually identical to the adjacent intestine.

Histological organization of the spiral valve wall represents a continuation of the adjacent intestine. However, the mucosal ridges become shorter distally and connective tissue gradually replaces the smooth muscle within the ridges. There is also greater vascularization of the wall and valve mucosa. The valves are composed of a connective tissue layer with strands of smooth muscle situated between two mucosal layers which are similar to that of the wall but thinner. Waves of mucus produced by goblet cells of the valves are exuded from the crypts. The typhosole consists of lymph nodules, connective tissue, and blood vessels.

The simple epithelium of the proximal rectum is dominated by goblet cells with a gradation to a stratified squamous layer immediately prior to the vent.

There is very little information concerning the histological characteristics of the accessory digestive organs of chondrosteans. General liver structure is similar to descriptions for other fish (Fange & Grove 1979) with the hepatocytes arranged into bilayered muralia separated by sinusoids which radiate from central vessels. Vacuoles within the hepatocytes contain lipid. This is enhanced in white sturgeon fed artificial diets containing 15–25% lipid. These levels are higher than those in natural foods of adults and juveniles and result in excessive hepatic fat deposition and possible liver dysfunction (Buddington & Doroshov 1984).

The diffuse pancreatic tissue requires histological identification. Acinar cells are grouped into rosettes which surround central canals that anastomose with larger ducts lined with a simple, cuboidal epithelium. Zymogen granules are revealed by the presence of eosinophilic, tryptophan positive bodies within the cells.

Splenic tissues consist of dense patches of hematoxylin positive cells and regions of connective tissue with scattered eosinophilic cells.

Digestive system development

Mortality of acipenserids is highest during embryonic and larval periods. As a result there has been greater interest in early developmental stages in efforts to understand how to increase survivorship and improve hatchery efficiency. Digestive system differentiation is virtually identical throughout the chondrosteans based on descriptions of various species by Neumayer (1930), Krayushkina (1957), Caloianu-Iordachel (1960, 1966), Ballard & Needham (1964), Afonich (1970), Schmal'gauzen (1972), Detlaf et al. (1981), and Buddington & Doroshov (1985c). Although development rate is temperature dependent and exhibits species specificity, hatching to first feeding usually requires 12–16 days at 17° C with metamorphosis of the larvae to juveniles occurring between 20 and 30 days. The following presents a generalized description of chondrostean digestive system development.

At hatching the alimentary canal is represented by the yolk endoderm and a partially differentiated hindgut which contains pigmented materials. Walls of the stomach and intestine, which will develop from the yolk endoderm, are syncytial layers of squamous cells. The mouth and accessory digestive organs are absent.

Two days after hatching a furrow begins to develop on the dorsal-posterior region of the yolk endoderm. It will eventually separate the intestine from the stomach. Differentiation of the alimentary canal is asynchronous, proceeding from distal to proximal. Corresponding with this, the spiral valve is partially differentiated by day three whereas the presumptive stomach and intestinal walls remain as syncytia. Hepatic tissue begins to form at this time as a single lobe on the ventral portion of the yolk endoderm.

The stomach and intestine are separated by days 6–8 and due to yolk absorption, especially from the midgut, the gut diameter decreases. Concurrently, there is a shift of the intestine to the right side of the body and its connection with the stomach moves anteriorly. Except for the ventral-posterior region, the rest of the stomach wall remains as a syncytium. A mass of tissue located in the esophagus begins to degenerate and thereby opens a connection between the mouth and stomach. By now the valves are fully formed in the spiral valve.

Developing gastric glands are present in the anterior stomach on day 10 and the pylorus is a region of intense smooth muscle proliferation, forming the future gizzard. Presumptive pancreatic tissue can be observed in the liver adjacent to the pylorus, however, zymogen granules are not yet present. When exogenous feeding is established on day 12 the digestive system is anatomically similar to the adult form. The yolk may or may not be exhausted when most species initiate exogenous nutrition with some larvae exhibiting a period of mixed feeding when both yolk and external food are present within the stomach. The pigment plug in the spiral valve is eliminated just prior to oral feeding or when the first feces are voided.

Between days 12 to metamorphosis the smooth muscle layers expand. In addition, there is a continued development of the mucosal epithelium in the post-gastric alimentary canal. During metamorphosis the swim bladder forms as an outgrowth of the forestomach and the characteristic adult body shape and behaviors are assumed.

Digestive physiology

Although digestive capabilities of the chondrosteans have not been investigated, preliminary research indicates they are probably similar to those of other fish. However, the digestive processes are unclear. Speculations of digestive mechanisms can be developed based on morphological features of the digestive system, enzyme complement, and nutrient uptake activities. Since gut structure is similar throughout the acipenserids the following description of the physiological characteristics of only a few species may apply for the group.

Chemical digestion is initiated in the forestomach by pepsin concentrations of 25–35 units per mg protein and an acidic environment in the range of 3–4 (Buddington 1985, Buddington & Doroshov 1985b). Both acid and pepsin are synthesized and secreted by oxyntopeptic cells of the gastric glands (Smit 1961). Other non-proteolytic enzymes usually associated with the post-gastric alimentary canal have also been detected in acipenserid stomachs (e.g. lipase and amylase; Buddington 1985, Buddington & Doroshov 1985a, b), however, their potential digestive contributions in the acidic environment is not known. The source of these enzymes is also unclear but could not be attributed to a dietary origin or a backflow of intestinal contents. Additional gastric enzymes have been reported for other fish (Fange & Grove 1979) but they may have had dietary sources. Absorption of amino acids and sugars, determined by an in vitro technique is low in the forestomach relative to the intestine and spiral valve (personal observations RKB). Gastric emptying is regulated by the pyloric sphincter such that only small amounts of the triturated feed from the gizzard enters the intestine at any one time. As a result intestinal contents of white sturgeon are less than 40% of gastric volumes (Buddington & Doroshov 1985b).

The post-gastric alimentary canal is the primary region of digestive and absorptive functions, particularly the spiral valve (Buddington 1983). Enzyme diversity and concentrations, and protein levels are higher relative to the stomach. There is an increase in pH to a slightly alkaline condition due, possibly, to pancreatic input of bicarbonate. Enzymes with alkaline pH optima predominate. Trypsin is the primary endoprotease with the corresponding exoprotease, carboxypeptidase B which hydrolyzes exposed peptides resulting from tryptic activity, also at high levels. In contrast, chymotrypsin and the associated carboxypeptidase A are present at low concentrations. These results suggest a cooperative relationship exists between the endo- and exoproteases of sturgeon as also observed in higher vertebrates (Barnard 1973). Similar to other fish, elastase and leucineaminopeptidase are present but at low levels (Buddington & Doroshov 1985b).

Lipase and amylase were detected throughout the gut of white sturgeon, including the stomach. However, the concentrations were low relative to terrestrial vertebrates fed similar diets. Since fat digestibility is high in all fish previously examined (Halver 1972), it is probably also high in sturgeon, but this remains to be investigated. Carbohydrates are generally poorly utilized by carnivorous salmonids whose natural feeds during phylogeny would contain negligible amounts of this material. Therefore, sturgeon, which are primarily carnivorous, may not be capable of utilizing high dietary levels of carbohydrate as an energy source to spare the protein requirement. This also merits further study.

Alkaline phosphatase activity was detected in only the intestine and spiral valve. Although its digestive contribution is unclear, it appears to be associated with nutrient transport processes in fish alimentary canals (Stroband et al. 1977). Levels of nutrient uptake observed in vitro coincided with alkaline phosphatase activity with highest amino acid and sugar transport in the spiral valve followed by the intestine of white sturgeon (personal observations RKB).

The digestive enzyme complements of acipenserids exhibit age related changes (Korzhuev & Sharkova 1967, Buddington 1985, Buddington & Doroshov 1985a). Three secretory phases are present and appear to correspond with life history events, the natural feeds of each interval, and possibly changing nutrient requirements (Fig. 1). Since the studies of Buddington (1985) and Buddington & Doroshov (1985a) utilized fish which were fed the same ration throughout, observed shifts in enzyme activities were age related, not dietarily induced, and probably represent a genotypic influence associated with development.

The first phase or yolksac interval coincides with digestive system differentiation. During this time enzyme concentrations are low within the gut and those that are present are most likely associated with intracellular enzymes of the yolk endoderm. Throughout this interval the embryos are depen-

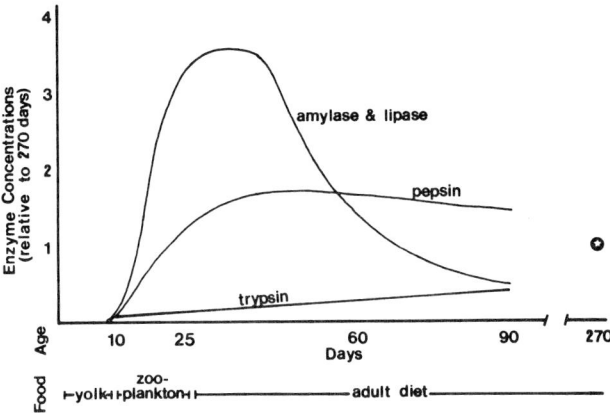

Fig. 1. Relative concentrations of digestive enzymes and food habits during development of acipenserids. From data of Korzhuev & Sharkova (1967), Buddington (1985) and Buddington & Doroshov (1985a, b). Enzyme concentrations (units per mg protein) at 270 days, indicated by the star: amylase 0.1; lipase 0.7; pepsin 26; trypsin 275.

dent on endogenous nutritional reserves and although utilization of the yolk is indicated by decreasing alimentary canal protein concentrations, mechanisms of uptake are unknown. To compensate for low enzyme activities, pinocytosis and intracellular digestion may be the primary processes of yolk utilization. This is substantiated by observations of yolk granules within vacoules of the developing alimentary canal epithelium of white sturgeon (Buddington 1983) and other acipenserids (Krayushkina 1957).

The second interval coincides with the larval period; from onset of active feeding to metamorphosis. When feeding is established enzyme concentrations and diversity increase. In addition to higher tryptic levels, proteolysis is supplemented by gastric secretion of acid and pepsin. In contrast to acipenserids, most fish develop gastric functions only after metamorphosis (Tanaka 1971). Amylolytic and lipolytic activities are highest during the acipenserid larval period which suggest carbohydrates and lipids may actually be more available to feeding larvae relative to older fish. This may reflect an adaptation to the larval zooplanktivorous diet which is dominated by cladocerans and copepods (Baranova & Miroshnichenko 1969, personal observations RKB). High lipolytic activity would enhance digestion of the lipids which are present in these organisms at levels up to 50% of the body weight (Tessier & Goulden 1982) and also permit utilization of less digestible wax esters (Patton et al. 1975) which can constitute significant proportions of zooplankton lipids (Lee et al. 1971). Since first feeding acipenserids possess a gut structure and enzyme complement which are similar to the adults, overall digestive capacities may be comparable among age groups. Digestive capabilities of feeding larvae of other fish, however, are likely to be lower relative to older fish due to incomplete gut structure and only a partial enzyme complement.

Following metamorphosis lipolytic and amylolytic levels decline whereas the protease concentrations continue to increase. This corresponds with the transition to the adult diet which, for the species investigated, contains lower lipid levels (Buddington 1985, Buddington & Doroshov 1985c).

Feeding characteristics and sensory adaptations

Fish utilize a variety of senses for recognition, orientation, location, and eventual acceptance of food items. Data available for benthophagous acipenserids indicate sight is not utilized during feeding (Sbikin 1974). Instead, sensitivities of other senses appear to have been intensified. Large olfactory rosettes, similar to those of other fish, are present (Hara 1971) and an abundance of taste buds are distributed on the four sensory barbels and on the ventral surface of the rostrum. Ampullary receptors have been described in the ventral rostrum of shovelnose sturgeon (Teeter et al. 1980) and other acipenserids (Norris 1925, Jorgenson 1972). The morphology and physiology of these receptors are similar to the Lorenzian ampullae of elasmobranchs and probably have electroreceptive functions. It remains to be ascertained whether sturgeon can detect the weak electrical currents generated by benthic organisms and thereby locate food on or within the substrate.

The responses of white sturgeon to chemical stimuli have been determined using extracts of live and manufactured feeds (unpublished data JPC). Initial studies utilizing the bioassay procedure of

Scherer & Nowak (1973) revealed fish fed live feeds were strongly attached to homogenates prepared from the feeds but exhibited little response to preparations of manufactured rations. Fish fed manufactured rations had comparatively lower responses to live feed extracts but their reaction to the formulated rations was not enhanced. Further work to characterize chemical stimulants present within tubifex and *Artemia* utilized homogenates of the organisms which were boiled, dialyzed, filtered through Diaflow filters with cutoffs at 500 and 1000 molecular weights, and exposed to 6N hydrochloric acid at 100°C; 24 hours. The response of white sturgeon indicated the stimulants are non-volatile, of low molecular weight, and are heat stable. Additional work is being performed to determine their solubility in alcohols and in non-polar solvents, and if they will bind to cation exchange resins. If so, the stimulants will coincide with the generalizations proposed by Carr (1982) and would be consistent with properties of free amino acids and other nitrogenous substances of low molecular weight.

The differing responses of juvenile sturgeon to live and formulated feeds implies the development of search images (Atema 1977), possibly based on the chemical stimulants present in the feeds. This may partially explain the imprinting phenomenon observed in feeding trials with acipenserids. Only a small percentage of sturgeon initially fed live feeds will eventually accept manufactured rations and attempts to wean the fish off live organisms result in high mortality (Charlon & Williot 1978, Monaco et al. 1981, personal observations RKB). Only a few days of live feed presentation are necessary for establishment of the search images for live feeds. The imprinting and later weaning problems have been avoided by exclusive use of formulated feeds. In contrast to the live organisms, sturgeon do not develop as strong an imprinting on manufactured rations and will still accept live feeds even after 100 days. However, their initial attempts to locate tubifex or other live feeds are not as effective as those of fish maintained on live organisms. This may have profound implications for stock replenishment operations.

The physico-chemical properties which influence acceptance of formulated feeds by sturgeon larvae have yet to be determined. Even though nutritional composition is similar among various manufactured larval rations, only a small number of these are accepted at first feeding. Consistency and texture may be of importance when preparing feeds for early life history stages. For example, hard, dry rations are poorly accepted at first feeding whereas softer, semi-moist formulations are ingested and promote high growth and survival (Buddington & Doroshov 1984). Observations have revealed onset of exogenous feeding is earlier when live feeds are fed, followed by semi-moist rations. There are substantial delays in first feeding followed by high mortality when dry rations are presented. It is possible though, the semi-moist formulations release more chemical stimulants and thereby increase feeding responses.

Feeding and development of formulated rations

In contrast to other intensively cultured fish of comparable commercial interest, there has been virtually no development or utilization of manufactured rations for sturgeon. As a result European acipenserids are raised using live feeds. During early development of sturgeon culture in the USSR moist artificial rations were formulated using bone and blood meals, silk worms, and mineral and vitamin supplements (Gordienko et al. 1970). Since such formulations were ineffective, the use of live organisms was adopted (Milstein 1972). Later attempts to utilize artificial rations also met with little success (Charlon & Williot 1978) and growout to market size was only possible using natural feeds (Suzuki & Nishi 1976, Romanycheva 1976). Recent work, though, by Barracund et al. (1979) and Monaco et al. (1981) demonstrated high growth potential of sturgeon fed artificial rations with weight gains greater than those of fish maintained on natural feeds. However, there was an excessive mortality associated with converting the fish from live organisms to the manufactured rations. This appeared to be due to the imprinting of the fish onto the live feeds, discussed previously.

Mortality associated with weaning lake and white sturgeon off live feeds has been avoided by presentation of a semi-moist salmonid formulation at initiation of feeding (Buddington & Doroshov 1984, personal observations, RKB). Although the manufactured rations initially elicited lower growth, weight gains of older fish maintained on the formulated feeds were higher relative to fish maintained on live feeds (Monaco et al. 1981). Dry formulations with nutritional compositions similar to the semi-moist rations were not well accepted at initiation of feeding and resulted in low larval growth and survival (Buddington & Doroshov 1984). However, once the fish have established feeding on the semi-moist feeds they can later be weaned to the less expensive, more convenient dry rations with virtually no increase in mortality. As a result, white sturgeon are now raised throughout the life cycle using manufactured rations exclusively (personal observations RKB). Additional investigations will be required to determine the suitability of manufactured rations for rearing of other acipenserids. If successful, such results will dramatically increase hatchery efficiency and decrease labor and feed costs.

Since existing manufactured feeds do not elicit feeding responses comparable to live organisms, they remain in the water for a considerable time during which leaching of the nutrients can occur. This problem is intensified with small particles that have high surface to volume ratios. Attempts to enhance feeding responses by incorporation of hydrolyzates prepared from brine shrimp (a preferred food) and other animal tissues were unsuccessful (unpublished data RKB). Efforts to decrease the nutrient leaching by addition of various binders also did not improve feed efficiency. Feeding trials did suggest texture as an important characteristic at first feeding as indicated by the contrast between feeding hard, dry rations and softer, semi-moist formulations. Electrical stimuli may also be involved. Live brine shrimp and tubifex elicit strong feeding responses but the reaction is diminished after freezing (personal observations (RKB). The fish may be detecting electrical currents generated by the organisms.

Efficient rations will also require elucidation of nutrient requirements, which are unknown for any acipenserid. Although empirical studies will be necessary, such work will be enhanced by application of data from feeding and physiological studies. The carnivorous habits and preponderance of proteases in the enzyme complement suggest protein will be the primary dietary constituent. Low amylase concentrations and low sugar transport indicate carbohydrates are not well utilized. This agrees with studies of other carnivorous fish (Palmer & Ryman 1972, Cowey & Sargent 1979, Buddington & Diamond 1985). Lipid requirements may change during the life history. Based on lipase data and feeding habits, optimal lipid levels for the larvae may be as high as 50% whereas older fish would require substantially lower quantities.

Conclusions

Digestive system structure seems to be similar throughout the chondrosteans. The alimentary canal exhibits an anatomical regionalization of functions and adaptations consistent with particular natural feeding habits. The associated physiological processes correspond with the group's carnivorous habits as indicated by the predominance of proteases and low concentrations of amylase. Digestive system development differs from other fish due to the presence of an intraembryonic yolk endoderm which participates in formation of the alimentary canal. In addition, the gut possesses a more advanced structure and function at first feeding relative to other fish.

Until recently, the lack of suitable manufactured rations and feeding techniques have been major impedances in the development of large scale, intensive growout of acipenserids. Although sturgeon can now be reared throughout the life cycle using manufactured feeds exclusively, the formulations now utilized were developed for salmonids. A better understanding of the feeding characteristics and the exploitation of such information would enhance the design of manufactured rations and feeding strategies. This would enhance feed acceptance, decrease leaching losses, and thereby increase feed efficiency. Nutritional requirements are unknown for any acipenserid, but will need to be quantified before effective formulations can be designed. Further research is also needed to determine the digestibility and suitability of potential feed components. Physiological data in conjunction with natural feeding habits suggest distinct formulations may be required for each life history interval.

References cited

Afonich, R. V. 1979. The feeding habits of stellate sturgeon larvae at the early stages of development in hatcheries. Trudy Vses. Nauchno-Issled. Inst. Morsk. Rybn. Khozy. Okeanograf. 74: 58–81. (In Russian).

Andrew, W. & C. P. Hickman. 1974. Histology of the vertebrates, a comparative text. A. V. Mosby Co., St. Louis. pp. 243–315.

Atema, J. 1977. Functional separation of smell and taste in fish and crustacea. pp 165–174. In: J. LeMagnen & P. MacLeod (ed.) Olfaction and Taste, Vol. 1, Information Retrieval, Washington, D.C.

Ballard, W. W. & R. G. Needham. 1964. Normal embryonic stages of Polyodon spathula (Walbaum). J. Morph. 114: 465–478.

Baranova, V. P. & M. P. Miroshnichenko. 1969. Conditions and prospects for culturing sturgeon fry in the Volgograd sturgeon nursery. Hydrobiol. J. 5: 63–67.

Barnard, E. A. 1973. Comparative biochemistry and physiology of digestion. pp. 133–164. In: C. L. Prosser (ed.) Comparative Animal Physiology, Saunders, Philadelphia.

Barracund, M., P. Ferlin, P. Lamarque & J. J. Sabaut. 1979. Alimentation artificielle de l'esturgeon (Acipenser baeri). pp. 411–422. In: J. E. Halver & K. Tiews (ed.) Finfish Nutrition and Fishfeed Technology, Vol. 1, Heeneman and Co., Berlin.

Buddington, R. K. 1983. Digestion and feeding of the white sturgeon, Acipenser transmontanus. Ph.D. Dissertation, University of California, Davis. 139 pp.

Buddington, R. K. 1985. Digestive secretions of lake sturgeon (Acipenser fulvescens) during early development. J. Fish Biol. (in press).

Buddington, R. K. & S. I. Doroshov. 1984. Feeding trials with hatchery produced white sturgeon juveniles (Acipenser transmontanus). Aquacult. 36: 237–243.

Buddington, R. K. & S. I. Doroshov. 1985a. Development of digestive secretions in white sturgeon (Acipenser transmontanus). Comp. Biochem. Physiol B. (in press).

Buddington, R. K. & S. I. Doroshov. 1985b. Digestive enzyme complement of white sturgeon (Acipenser transmontanus). Trans. Amer. Fish. Soc. (in press).

Buddington, R. K. & S. I. Doroshov. 1985c. Anatomy and histology of the white sturgeon digestive system (Acipenser transmontanus). J. Fish Biol. (in press).

Buddington, R. K. & J. M. Diamond. 1985. Diet dependence of vertebrate intestinal nutrient transport: a genotypic component. Fed. Proc. 44: 811.

Caloianu-Iordachel, M. 1960. Contribution to the histophysiological study of the digestive apparatus in the Black Sea sturgeon, star sturgeon, great sturgeon, and in Acipenser nudiventris Lov. during the first stages of post-embryongensis. Rev. Biol. 5: 319–335. (In Roumanian).

Caloianu-Iordachel, M. 1966. Characteristic formation of the digestive tract in Acipenseridae in comparison with other lower vertebrates. Hidrobiologia 7: 81–94. (In Roumanian).

Carr, W. E. S. 1982. Chemical stimulation of feeding behavior. pp. 259–273. In: T. J. Hara (ed.) Chemoreception of Fishes, Vol. 8, Elsevier Scientific Publ., New York.

Charlon, N. & P. Williot. 1978. Rapport sur l'elevage des esturgeon en Union Sovietique. Ministere de l'Agriculture, CTGREF-INRA: 42–45.

Cowey, C. B. & J. R. Sargent. 1979. Nutrition. pp. 1–70. In: W. S. Hoar, D. J. Randall & J. R. Brett (ed.) Fish Physiology, Vol. 8, Academic Press, New York.

Detlaf, T. A., A. S. Ginzburg & O. I. Schmal'gauzen. 1981. Development of the sturgeon fishes. Nauka Press, Moscow. 224 pp. (In Russian).

Fange, R. & D. Grove. 1979. Digestion. pp. 162–260. In: W. S. Hoar, D. J. Randall & J. R. Brett (ed.) Fish Physiology, Vol. 8, Academic Press, New York.

Golovanenko, L. F. 1964. The physiological condition of the young sturgeon reared on different feeds. Izv. Gos. Nauchno-Issled. Inst. Ozern. Rechn. Ryb. Khoz. 57: 235–241. (In Russian).

Gordienko, O. L., R. V. Afonich & E. V. Soldatova. 1970. The rearing of the young of the sturgeon fishes on artificial foods in tanks. Trudy Vses. Nauchno-Issled. Inst. Morsk. Rybn. Khoz. Okeanogr. 74: 7–36. (In Russian).

Halver, J. E. 1972. Fish nutrition. Academic Press, New York. 714 pp.

Hara, T. J. 1982. Chemoreception in fishes. Elsevier Scientific Publ., New York. 295 pp.

Hopkins, C. S. 1895. On the enteron of American ganoids. J. Morph. 11: 411–422.

Jorgenson, J. M., A. Flock & J. Wersall. 1967. The Lorenzian ampullae of Polyodon spathula. Z. Zellforsch. 130: 362–372.

Korzhuev, P. A. & L. B. Sharkova. 1967. Digestion characteristics of Caspian sturgeon. pp. 326–330. In: G.S. Karzinkin (ed.) Metabolism and Biochemistry of Fishes, Nauko Press, Moscow. (In Russian).

Krayushkina, L. S. 1957. The histology of the digestive system of sturgeon larvae at different developmental stages. Dokl. Akad. Nauk. SSSR 177: 966–968. (In Russian).

Lee, R. F., J. Hirota & A. M. Barnett. 1971. Distribution and importance of wax esters in marine copepods and other zooplankton. Deep-Sea Res. 18: 1147–1165.

Macallum, A. B. 1886. The alimentary canal and pancreas of Acipenser, Amia, and Lepidosteus. J. Anat. Physiol. 20: 604–636.

Magid, A. M. A. 1975. The epithelium of the gastrointestinal tract of Polypterus senegalus (Pisces: Brachiopterygii). J. Morph. 146: 447–456.

Milstein, V. V. 1972. Sturgeon culture. Pishch. Prom., Moscow.

129 pp. (In Russian).

Monaco, G., R. K. Buddington & S. I. Doroshov. 1981. Growth of white sturgeon under hatchery conditions. Proc. 12th Annu. Meet. World Maricult. Soc. 12: 113–121.

Neumayer, L. 1930. Die Entwicklung des Darms von *Acipenser*. Acta Zool. 11: 1–114.

Norris, H. W. 1925. Observations upon the peripheral distribution of the cranial nerves of certain ganoid fishes (*Amia, Lepidosteus, Polyodon, Scaphyrhynchus,* and *Acipenser*). J. Comp. Neurol. 39: 345–416.

Palmer, T. N. & B. E. Ryman. 1972. Studies on oral glucose intolerance in fish. J. Fish Biol. 4: 311–319.

Patton, J. S., J. C. Nevenzel & A. A. Benson. 1975. Specificity of digestive lipases in hydrolysis of wax esters and triglycerides studies in anchovy and other selected fish. Lipids 10: 575–583.

Romanycheva, O. D. 1976. The modern condition of the development of the marine sturgeon culture for commercial purposes and its perspectives. Proc. 5th Japan-Soviet Joint Symp. Aquacult., Tokai University, Tokyo: 353–365.

Sbikin, Yu. N. 1974. Age related changes in the role of vision in the feeding of various fish. J. Ichthyol. 14: 133–138.

Scherer, E. & S. Nowak. 1973. An apparatus for recording avoidance movements of fish. J. Fish. Res. Board. Can. 30: 1954–1956.

Schmal'gauzen, O. I. 1972. Effects of phenol on development of the prolarvae of acipenserid fish III. The digestive system. Sov. J. Dev. Biol. 3: 407–413.

Smit, H. 1961. Gastric secretion in the lower vertebrates and birds. pp. 2791–2805. *In:* C. F. Code (ed.) Handbook of Physiology, Section 6, Alimentary Canal, Vol. 6, Chapter 135, American Physiological Soc., Washington D. C.

Stroband, H. W. J., H. v.d. Meer & L. P. M. Timmermans. 1979. Regional functional differentiation in the gut of the grasscarp, *Ctenopharyngodon idella* (Val.). Histochem. 64: 235–249.

Suzuki, K. & G. Nishi. 1976. Rearing of sturgeons in Japan. Proc. 5th Japan-Soviet Joint Symp. on Aquacult., Tokai University, Tokyo: 379–393.

Tanaka, M. 1971. Studies on the structure and function of the digestive system in teleost larvae-III. Development of the digestive system during post-larval stages. Jap. J. Ichthyol. 18: 164–174.

Teeter, J. H., R. B. Szamier & M. V. L. Bennett. 1980. Ampullary electroreceptors in the sturgeon *Scaphyrhynchus platorhyncus* (Rafinesque). J. Comp. Physiol. 138: 213–223.

Tessier, A. J. & L. E. Goulden. 1982. Estimating food limitations in cladoceran populations. Limnol. Oceanogr. 27: 707–717.

Weisel, G. F. 1973. Anatomy and histology of the digestive organs of the paddlefish (*Polyodon spathula*). J. Morph. 140: 243–256.

Weisel, G. F. 1979. Histology of the feeding and digestive organs of the shovelnose sturgeon, *Scaphyrhynchus platorhyncus*. Copeia 1979: 518–525.

Received 15.5.1984   Accepted 25.3.1985

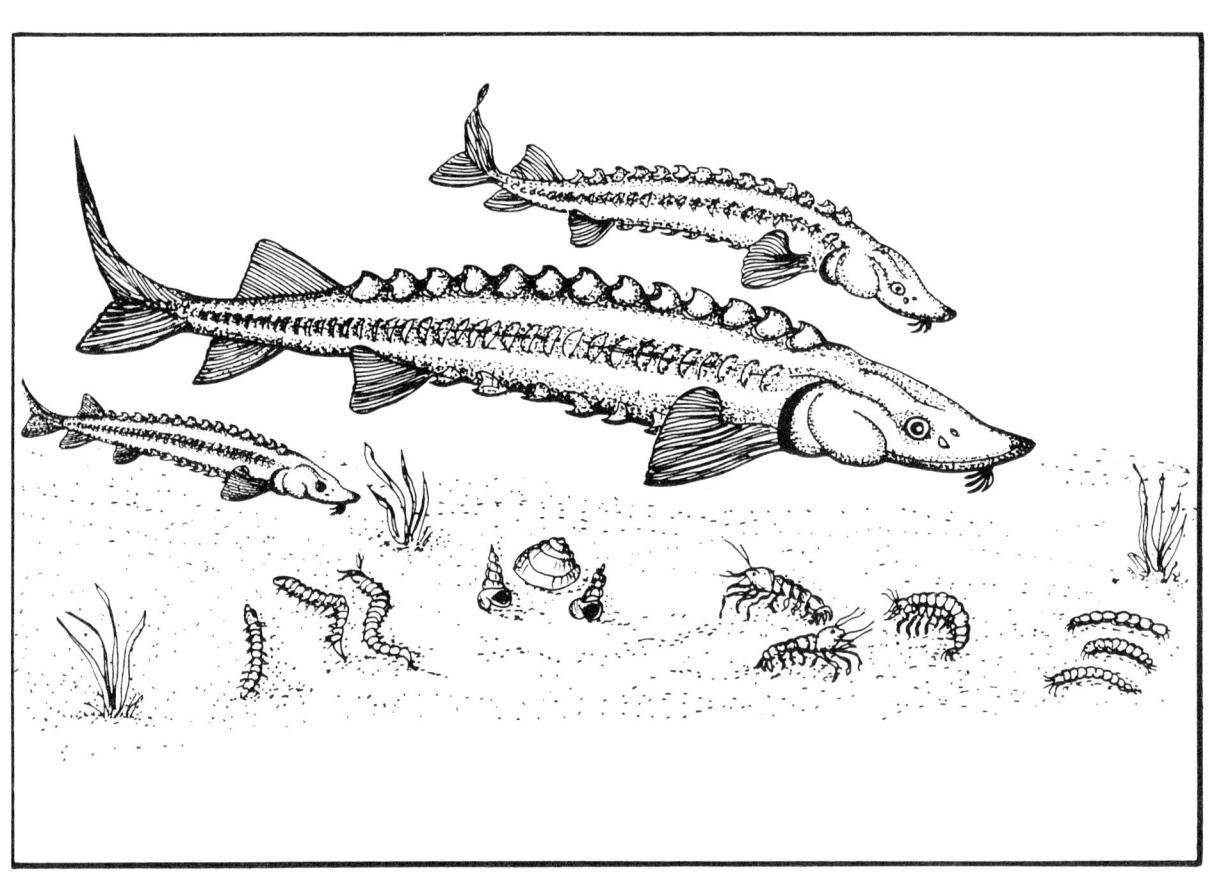

Effect of temperature on early development of white and lake sturgeon, *Acipenser transmontanus* and *A. fulvescens*

Yuan L. Wang[1], Frederick P. Binkowski[2] & Sergei I. Doroshov[1]
[1] *Department of Animal Science, University of California, Davis, CA 95616, U.S.A.*
[2] *Center for Great Lakes Studies, University of Wisconsin-Milwaukee, Milwaukee, WI 53201, U.S.A.*

Keywords: Chondrostean, Acipenserid, Embryos, Larvae, Rate of development, Survival, Temperature adaptations

Synopsis

The effect of constant incubation temperatures (between 10°C and 26°C) on the developmental rates was found to fit a similar exponential relationship in both the lake and white sturgeon embryos and larvae. Although the lake sturgeon had an overall slower rate of development than the white sturgeon, no statistically significant difference was detected in the slopes of the exponential equations describing the effect of temperature on developmental rate. The effect of these incubation temperatures on embryonic survival also did not differ between these two species. Both species exhibited optimal survival between 14–17°C and incipient mortalities occurred at 20°C. Temperatures above 20°C were lethal for white sturgeon embryos. No effect of low incubation temperature on survival was evident from this study. A comparison of these North American species with Eurasian acipenserids suggests that all the sturgeon that have been examined exhibit a similar influence of incubation temperature on developmental rate.

Introduction

We examine the effect of temperature on the timing of early development and survival of two North American Acipenserid species: white sturgeon, *Acipenser transmontanus*, from San Francisco Bay, and lake sturgeon, *Acipenser fulvescens*, from the Great Lakes. These two species were abundant and commercially important in the past century but their numbers and harvest have dramatically declined due to overharvesting, pollution and dam construction (Harkness & Dymond 1961, Miller 1972). Recent investigations have shown that temperature is an important factor in the success of the artificial propagation of sturgeon (Binkowski & Czeskleba 1980, Doroshov et al. 1983). Understanding the environmental requirements, specifically temperature and how it effects development, growth and survival during the early life history stages, is essential for the successful hatchery management of sturgeon.

Detlaf et al. (1981) investigated the effect of temperature on cleavage and organogenesis in three acipenserid species from the Caspian and Azov seas. No similar information is available for North American sturgeon species. Normal stages of early development were described at ambient temperature (14 to 20°C) for the Atlantic sturgeon (Dean 1895), lake sturgeon (Harkness & Dymond 1961), white sturgeon (Beer 1980) and paddlefish (Ballard & Needham 1964).

* Contribution No. 277 Center for Great Lakes Studies, University of Wisconsin-Milwaukee, Milwaukee, WI 53201, USA.

Material and methods

The observations were conducted in spring 1983 at the University of Wisconsin-Milwaukee (lake sturgeon) and at the University of California-Davis (white sturgeon). Zygotes of white sturgeon were obtained from a single mating. Broodfish were caught during the spawning run in the Sacramento River, then induced to ovulate by the administration of common carp pituitary extracts. Eggs of lake sturgeon were collected from two naturally ovulated females captured on spawning grounds below Eureka Dam on the Fox River, Wisconsin. Gamete collection, insemination and egg deadhesion procedures were similar for both species (Doroshov et al. 1983), and carried out in fresh water of 13–15°C.

The fertilized eggs and embryos which emerged were incubated at different constant (± 0.5°C) temperatures until yolk absorption. Post fertilization time and survival at the developmental stages described below, were determined for both species.

Experimental procedures differed for lake and white sturgeon, and data obtained should be treated accordingly. Six temperature treatments: 11°C, 14°C, 17°C, 20°C, 23°C, 26°C, with four replications, were applied for white sturgeon. Approximately 1000 eggs in each replicate were incubated in conical plastic jars with an upwelling water flow. Jars were installed inside 15 liter round tanks, supplied with fresh ground water in six independent temperature controlled blocks. Emerging embryos escaped via outflows from the jars into the tanks and remained there until the end of the experiment. Eggs of lake sturgeon were incubated 'en masse' in commercial MacDonald jars at constant temperatures 10°C, 15°C, and 20°C, without replication. Embryos that emerged were held in insulated fiberglass tanks supplied with water of the same temperature. In both cases, water was aerated and dissolved oxygen was maintained at saturation.

We attempted to sample the embryos and larvae of both species at certain developmental stages, classified by Detlaf & Ginzburg (1954), as follows: stage 6 – third cleavage at animal pole; 14 – horizontal blastopore furrow; 22 – closure of neural tube; 29 – differentiation of S-shaped heart; 35 – initiation of hatching; 36 – completion of hatching; 40 – separation of intestine and stomach; 44 – completion of yolksac absorption, discharge of pigment plug from spiral valve. These stages are described in the 'Results' section.

Development of white sturgeon was monitored by 'in vivo' microscopic examination of 15–20 animals at 1 to 12 hour intervals, depending on stage and treatment. When the proportion of animals reaching the above stages was ≥ 50% of total sample, the post fertilization time was recorded and 50 to 80 specimens were sampled from each replicate and preserved in buffered formalin.

'In vivo' monitoring of lake sturgeon was not done. Instead, lake sturgeon were sampled at frequent intervals (every 4 hours for embryos before and every 20 hours after hatching, preserved in buffered formalin and later examined. The samples which contained >50% of the sturgeon at stages 6 and 40 were missing from one or more temperature treatments and these stages were deleted from the final data analysis.

Preserved specimens were examined and photographed under a dissecting microscope and the percentage of normally developing animals was counted in each sample. This was assumed to express the survival prior to hatching (all embryos that emerged survived from hatching to the end of the experiments). Data were pooled for the stages 6–14 (cleavage to gastrulation) and stages 22 through 36 (neurulation to hatching), since no difference in proportions of normally developing embryos was observed within these two groups.

The relationships between the timing of development and temperature were computed for each stage using exponential equations. The relationship between survival and temperature was fitted by least squares regression. All regressions yielded significant F values (≤ 0.05) and R^2 values ranged from 0.935 to 0.999.

Results

Embryonic development

Developmental stages, used for sampling during these experiments, are illustrated by photographs (preserved specimen) in Figure 1 and 2. Zygotes of white sturgeon are 4 mm in diameter and darkly pigmented with melanine. Eggs of lake sturgeon are smaller (3.5 mm) and lighter in color (brownish grey). Eggs of both species exhibit identical patterns of holoblastic cleavage. At the stage of third cleavage (CLVG, #6), the first cleavage furrow completely divides the egg. The second cleavage furrow traverses slightly beyond the equator of the egg, and the two third cleavage furrows are restricted to the animal hemisphere. The latter consists of eight micromeres, slightly unequal in shape and size. Bright white coloration retains the configuration of the faint grey crescent, present in both species prior to and during early cleavage stages.

At gastrulation the embryonic development of both species proceeds in similar fashion. The dorsal lip of the blastopore can first be observed as a faint pigmented line appearing slightly above the equator. Involution progresses as the lip elongates into

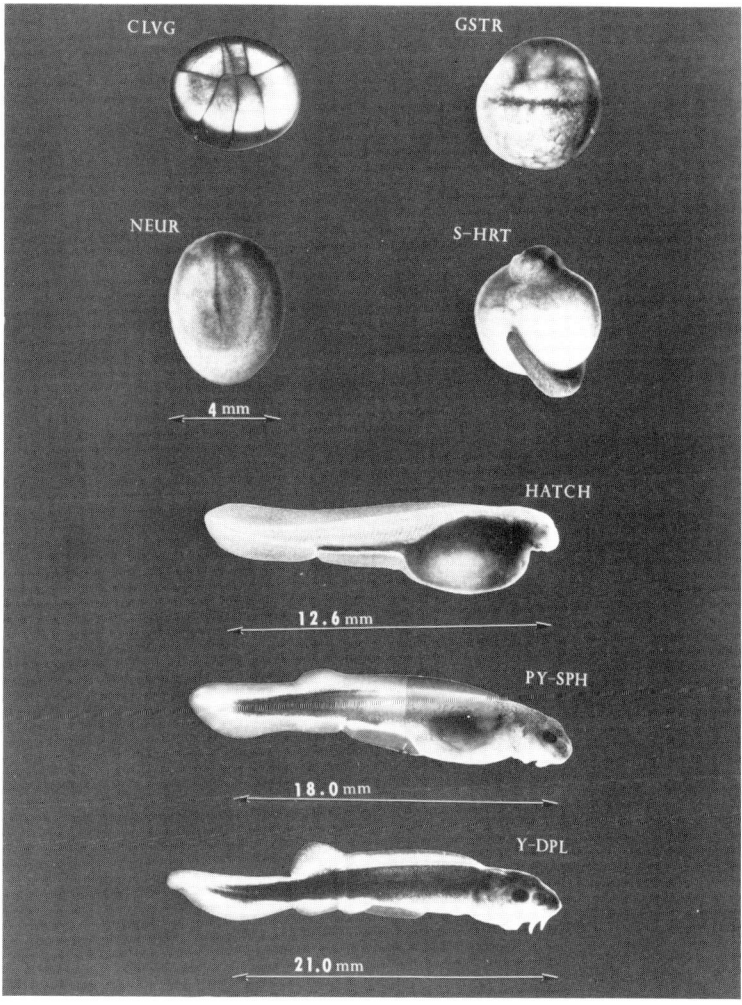

Fig. 1. Stages of embryonic development of white sturgeon, *Acipenser transmontanus:* CLVG – early cleavage, 8 micromeres (6); GSTR – early gastrula, horizontal blastopore (14); NEUR – late neurula, closure of neural tube (22); S-HRT – differentiation of S-shaped heart (29); HATCH – emerged embryo (35 and 36); PY-SPH – separation of intestine and stomach, differentiation of pyloric sphincter (40); Y-DPL – completion of yolk absorption (44).

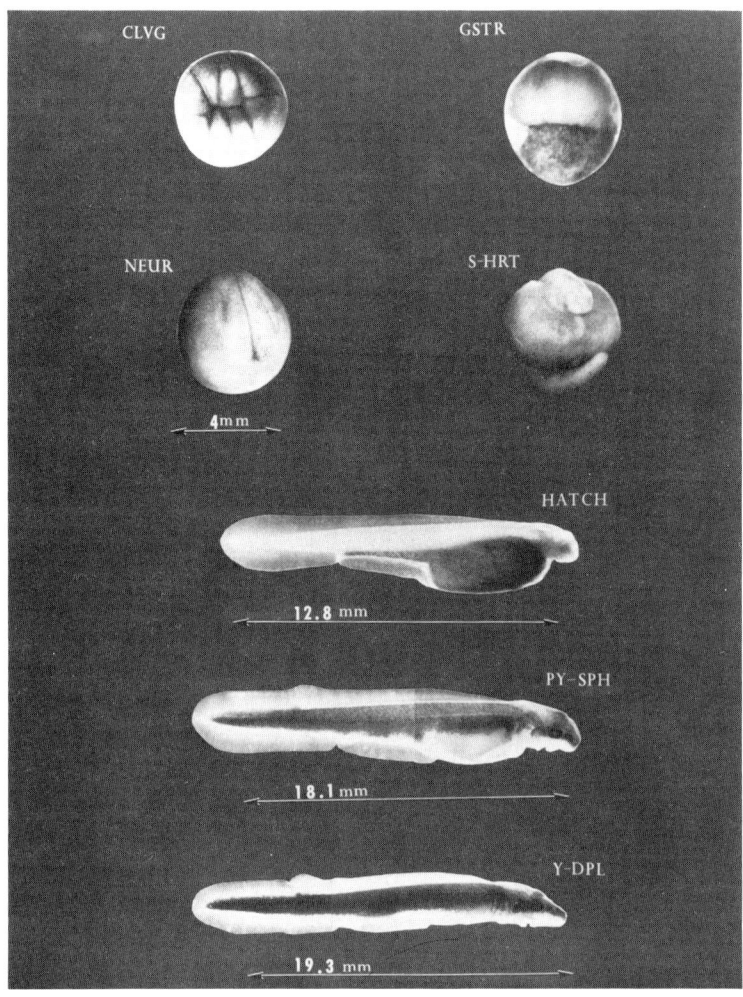

Fig. 2. Stages of embryonic development of lake sturgeon, *Acipenser fulvescens*. See Figure 1 for abbreviations.

the horizontal furrow (GSTR, #4) and the ectodermal cell mass moves toward the vegetal pole. Light or almost white colored blastoderm later surrounds dark endodermal cells resulting in the 'blastopore plug' stage, similar in both species. Before blastopore closure, the faint neural plate with pigmented folds extends from near the blastopore to almost half way around the egg. The neural folds move towards and join each other to form the neural tube. Neural fold closure first occurs at the mid section of embryo and then extends toward the cephalic region and toward the tail (NEUR, #22). The entire region is elevated and the first somites can be seen during the final period of neurulation.

The primordial heart appears as a straight tube projection below the head region of the embryo at the completion of neural tube closure. Following completion of neurulation, the heart tube flexes, assuming an 'S' shaped form. At this stage, the head region is elevated and the tail region is separated from the yolksac (S-HRT, #29).

As organogenesis continues, the tail elongates beyond the head region and the embryo begins to move. A pink colored substance (apparently associated with a release of hatching enzyme) appears in the perivitelline space a few hours prior to hatching. Hatching continues from one to several days, depending on water temperature. Embryos of white and lake sturgeon, collected at the beginning and completion of hatching (HATCH, #35 and 36) were similar in their appearance and size (10 to 11 mm, total length). Newly emerged embryos of lake sturgeon have light brown coloration and a more elongated yolksac. White sturgeon embryos have dark pigmentation and an almost round, dark colored yolksac.

The latter in sturgeon is composed of yolk endoderm which forms the gastrointestinal tract. The posterior (intestinal) portion of yolksac is depleted first. As the intestinal region of yolksac reduces in size, its color changes from greyish (in white sturgeon) or brown (in lake sturgeon) to white and a narrow connection, the primordial pyloric sphincter, clearly divides the yolksac into two regions: the intestine, void of yolk material (with an exception of oil droplets), and the future gastric region, still filled with dark yolk (PY-SPH, #40). Differences in body shape, finfold differentiation and pigmentation are now apparent between the two species. Lake sturgeon have a more streamlined configuration, an undifferentiated finfold and a faint longitudinal band of melanophores extending from snout to caudal region. White sturgeon appear more robust, their dorsal, caudal and pelvic fins are differentiating, a darkly pigmented band extends through only the caudal region, from slightly behind the anus to almost the end of the notochord.

As development advances, the yolk material eventually becomes depleted and replaced with the differentiated gastric region. The gut region is retracted and streamlined with the rest of the body. The pigmented 'melanine plug' appears in the spiral intestine and is soon discharged through the anus prior to the initiation of exogenous feeding (Y-DPL, #44). At this point white and lake sturgeon are clearly distinct in their appearance and morphological features; in white sturgeon all fins are differentiated, including the heterocercal caudal fin, and the dorsal row of scutes is differentiating. Metamorphic processes in lake sturgeon are delayed. Instead, larvae of this species assume the appearance of a burrowing form; the finfold is not differentiated, the body is elongated and the snout points downwards. Darkly pigmented bands extend along both sides of the body, from the tip of snout to the end of the notochord.

Timing of the development

Post-fertilization times required to reach the above described stages at the experimental temperatures are given in Table 1. Approximately two-fold acceleration in the development is evident as early as at gastrulation (stage #14) in both species, at temperatures $10°$ to $20°C$. Acceleration further increases in more advanced stages, especially in lake sturgeon. Lake sturgeon development appears to be slower, compared with that of white sturgeon, particularly in the lower temperature range. Embryos of white sturgeon emerge at 230–311 hours post-fertilization at $11°C$ and at 84–98 hours at $20°C$. Hatching of the lake sturgeon was observed at 380–430 hours at $10°C$ and 90–105 hours at $20°C$. White sturgeon reached the yolk depletion stage at 708 hours post-fertilization at $11°C$, while lake sturgeon reached a similar stage at 1316 hours post-fertilization at $10°C$.

Exponential equations describing the relationship between developmental time and temperature are given in Table 2. Slopes of the exponential lines (term 'b') tend to become more negative in more advanced embryonic stages, except for the white sturgeon late stages (Table 2). We found no statistically significant difference in terms 'b' among all stages of white and lake sturgeons, except for the last stage #44 'b' term was significantly more negative in lake sturgeon, $P<0.05$). However, all 'a' coefficients, characterizing elevation of curves, were significantly ($P<0.05$ to <0.001) greater in lake sturgeon, reflecting a slower rate of development in this species.

Survival curves were similar for both species (Fig. 3). All embryonic mortalities occurred between the early gastrula and late neurula. Range of temperatures between $12°C$ and $16°C$ appear to be optimal for embryonic survival in both species. Higher temperatures ranging from $20–22°C$ for embryos during cleavage and from $18–20°C$ for embryos during organogenesis showed lowered survival. The low temperature tolerance limits for both species are, apparently, below $10°C$ and were not evident in this study.

Table 1. Times required to reach various developmental stages and observed survival of white and lake sturgeon at the experimental temperature ranges.

White sturgeon							Lake sturgeon		
Temperature °C	11	14	17	20	23	26	10	15	20
	Time (hours after fertilization)								
Stages									
6	8	7	6	6	5	–	–	–	–
14	37	34	20	19	16	–	51	36	24
22	97	69	44	36	–	–	118	75	40
29	116	97	73	59	–	–	190	99	65
35	230	165	112	84	–	–	380	164	90
36	311	186	131	98	–	–	430	215	105
40	544	380	279	206	–	–	–	–	–
44	708	563	469	371	–	–	1316	524	308
	Proportions of normally developing embryos (% in samples).								
6–14	95.1	97.3	95.1	93.8	9.9	0	86.7	95.2	74.4
	(0.8)*	(1.4)	(1.0)	(0.9)	(4.3)				
22–36	87.6	88.6	83.6	49.1	0	0	67.0	74.9	47.4
	(1.5)	(2.2)	(1.9)	(3.2)					

* Numbers in parentheses are standard errors.

Table 2. Relationship between the timing of embryonic stages (Y-hours after fertilization) and incubation temperature (T) for lake and white sturgeon ($Y = ae^{bT}$).

Stages	White sturgeon		Lake sturgeon	
	a	b	a	b
6	11.33	−0.034	–	–
14	97.18	−0.084	109.00	−0.075
22	334.60	−0.114	358.43	−0.108
29	275.56	−0.077	534.39	−0.107
35	800.13	−0.114	1541.47	−0.144
36	1185.03	−0.127	1767.87	−0.141
40	1744.71	−0.107	–	–
44	1535.62	−0.071	5269.51	−0.145

Discussion

These results should be helpful in defining proper temperature conditions required for successful egg incubation and larval rearing of white and lake sturgeons. Embryos of both species exhibit similar temperature requirements for their normal development and survival. Successful egg incubation is possible within the temperature range 10°C to 18°C, but best results (highest survival and uniform hatching appear to be expected within the relatively narrow range of 14° to 16°C. Temperatures of 18° to 20°C may cause substantial mortalities during the sensitive embryonic stages, and temperatures above 20°C are clearly lethal, at least for the embryos of white sturgeon. Lower temperatures (10° and 11°C) exert an insignificant effect on embryonic mortality, and sublethal low temperatures are, apparently, below 10°C for both species. However, the application of temperatures below 14°C greatly

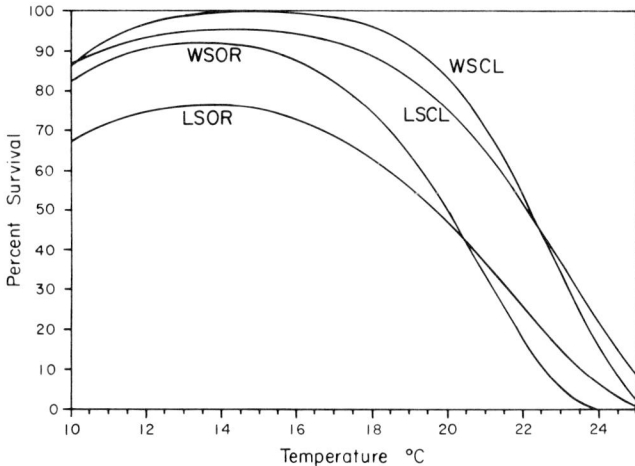

Fig. 3. The relationships between the proportions of normally developing embryos (survival) and incubation temperature: WSCL and WSOR – white sturgeon embryos during cleavage and gastrulation (stages 6–14), and organogenesis and hatching (22–36); LSCL and LSOR – lake sturgeon embryos at similar stages.

extended duration of incubation and hatching time in both species. This is undesirable in normal hatchery practice, due to increasing risk of higher mortality caused by fungal growth, a critical factor determining production of juveniles in sturgeon culture.

Our preliminary data show that embryos of white and lake sturgeon exhibit substantial similarity in their adaptations to environmental temperature. The overall effect of water temperature on the rates of development was quite similar in these allopatric species, although the lake sturgeon exhibited slower development at comparable temperatures. Both species had identical range of temperatures favorable for their survival. Experimental data are in agreement with field observations on spawning of these species. Harkness & Dymond (1961) indicated a temperature range 14–16°C as optimal for spawning of lake sturgeon. Kohlhorst (1976) observed the peak of white sturgeon spawning in the Sacramento River at a temperature of 14.4°C.

Reproduction in fishes is seasonal and their embryos and larvae exhibit species-specific and relatively narrow optimal temperature ranges (Blaxter 1969, Alderdice & Forrester 1971). Given the wide geographic range of the sturgeon species, we would expect substantial adaptive radiation in spawning temperature. Data for Eurasian species, reviewed by Detlaf et al. (1981), and data of this study appear to show the opposite, i.e. all sturgeon species studied appear to be conservative in their adaptations to spawning temperature. Sterlet, *Acipenser ruthenus*, beluga, *Huso huso*, Russian sturgeon, *Acipenser güldenstädti* and even Siberian sturgeon from the Lena River, *Acipenser baeri*, exhibit the same range of spawning temperatures of 10°C to 18°C (Schmidtov 1939, Detlaf & Ginzburg 1954, Detlaf 1970, Igumnova 1975, Nikolskaya & Sytina, 1978). The only exception is, perhaps stellate sturgeon, *Acipenser stellatus*, adapted to spawn in slightly warmer water, 15 to 25°C. Detlaf and co-workers provided experimental evidence that upper tolerance limits for embryos in all species, except for sevrjuga, is 20°C. These researchers carefully state that separation between sturgeon species by their adaptations to reproduction at different temperatures has not yet evolved (Detlaf et al. 1981).

The curves on Figure 4 show the relationships between duration of embryonic development (to hatching stage 35) and incubation temperature in five sturgeon species. Equations for the beluga, Russian and stellate sturgeon were computed from published data (Detlaf et al. 1981). The curves for white sturgeon, Russian sturgeon and stellate

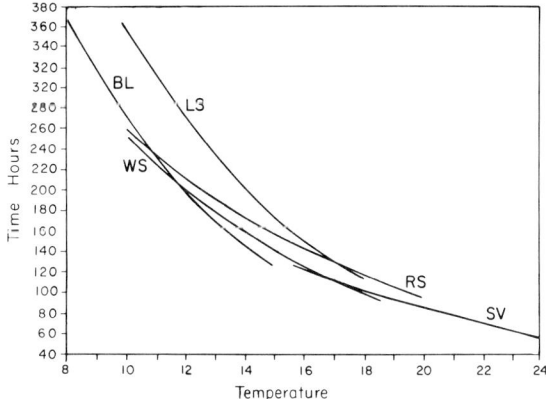

Fig. 4. Effect of temperature on the duration of embryonic development (to hatching, stage 35) in five sturgeon species: LS – lake sturgeon; WS – white sturgeon; RS – Russian sturgeon; BL – beluga; SV – stellate sturgeon. Data for RS, BL and SV are from Detlaf et al. (1981).

sturgeon are practically identical. Those for beluga and lake sturgeon have similar slopes but differ in elevation.

Acipenserids differ from modern teleosts in their possession of holoblastic cleavage and the morphogenetic processes governing their organogenesis (Detlaf & Ginzburg 1954). Ballard & Ginzburg (1980) pointed out the extreme similarity in major patterns of embryonic development existing among all sturgeon species, as a result of conservative holoblastic 'style' of early development. The effect of temperature on rate of mitosis during the early cleavage was found to be similar in several sturgeon species and was used as a universal 'dimensionless' characteristic to express the timing of early development (Detlaf & Detlaf 1960, Detlaf et al. 1981). It is possible that all sturgeon species, being highly conservative in their embryogenesis, respond with grossly similar metabolic processes to fluctuating incubation temperatures. Their embryos and larvae appear to tolerate a wide range of temperature during the reproductive seasons, roughly between 10 and 20° C in all species, although the optimal ranges may differ in different species and, possibly, in different stocks of one species.

Acknowledgements

Kenneth E. Beer (The Fishery, Sacramento) made this work possible by donating fertilized white sturgeon eggs for our study. We also thank the Department of Natural Resources, Wisconsin, and the California Department of Fish and Game for their continuous support of our research on sturgeon. This work was funded [in part] by the University of Wisconsin Sea Grant Institute under grants from the National Sea Grant College Program, National Oceanic and Atmospheric Administration, U.S. Department of Commerce, and from the State of Wisconsin. Federal grant NA80AA-D-00086, project R/LR-26.

References cited

Alderdice, D.F. & C.R. Forrester. 1971. Effect of salinity and temperature on embryonic development of petrale sole *Eopsetta jordani*. J. Fish. Res. Board Can. 28: 727–744.

Ballard, W.W. & A.S. Ginzburg. 1980. Morphogenetic movements of acipenserid embryos. J. Exp. Zool. 213: 69–103.

Ballard, W.W. & R.G. Needham. 1964. Normal embryonic stages of *Polyodon spathula* (Walbaum). J. Morph. 114: 465–478.

Blaxter, J.H.S. 1969. Development: eggs and larvae. pp. 177–252. In: W.S. Hoar & D.J. Randall (ed.) Fish Physiology, Vol. 3, Academic Press, New York.

Beer, K.E. 1980. Embryonic and larval development of *Acipenser transmontanus*. M.S. Thesis, University of California-Davis, Davis. 93 pp.

Binkowski, F.P. & D.G. Czeskleba. 1980. Methods and techniques for collecting and culturing lake sturgeon eggs and larvae. Abstract of paper presented in the annual meeting of The World Mariculture Society, New Orleans.

Dean, B. 1895. The early development of gar-pike and sturgeon. J. Morph. 11: 1–62.

Detlaf, T.A. 1970. The effect of environmental temperature during the oocyte maturation and ovulation on quality of sturgeon eggs obtained in the hatchery. Trudy Inst. Osetr. Khoz. 2: 112–126. (In Russian).

Detlaf, T.A. & A.A. Detlaf. 1960. On the dimensionless characteristic of the duration of embryonic development. Dokl. Akad. Nauk. SSSR 134: 199–202. (In Russian).

Detlaf, T.A. & T.A. Ginzburg. 1954. Embryonic development in sturgeon, in connection with problems of artificial propagation. Nauka Publishers, Moscow. 204 pp.

Detlaf, T.A., T.A. Ginzburg & O.I. Shmal'gauzen. 1981. Development of sturgeon: egg maturation, fertilization, embryonic and prelarval development. Nauka Publishers, Moscow. 224 pp.

Doroshov, S.E., W.H. Clark, P.B. Lutes, R.L. Swallow, K.E. Beer, A.B. McGuire & M.D. Cochran. 1983. Artificial propagation of white sturgeon *Acipenser transmontanus* (Richardson). Aquaculture 32: 93–104.

Harkness, W.J.K. & J.R. Dymond. 1961. The lake sturgeon: the history and problems of conservation. Fish and Wildlife Branch, Ontario Dept. of Land and Forest, Toronto. 121 pp.

Igumnova, L.V. 1975. Chronological patterns of embryonic development of beluga. The Soviet J. of Dev. Biol. 6: 38–43.

Kohlhorst, D.W. 1976. Sturgeon spawning in the Sacramento River in 1963 as determined by distribution of larvae. Calif. Fish and Game 62: 32–40.

Miller, L.W. 1972. White sturgeon population characteristics in the Sacramento-San Joaquin estuary as measured by tagging. Calif. Fish and Game 58: 94–101.

Nikol'skaya, N.G. & L.A. Sytina. 1978. A comparative analysis of constant temperatures on the embryonic development of four species of sturgeon. J. Ichthyology 18: 86–98.

Schmidtov, A.I. 1939. The sterlet *Acipenser ruthenus* L. Uchenie Zapiski, Kazanaski Gosud. Univ. Zoologia 99 (6/78): 1–279.

Received 15.5.1984   Accepted 18.3.1985

Distribution, biology and hybridization of *Scaphirhynchus albus* and *S. platorynchus* in the Missouri and Mississippi rivers

Douglas M. Carlson[1], William L. Pflieger, Linden Trial & Pamela S. Haverland
Missouri Department of Conservation, 1110 College Avenue, Columbia, MO 65201, U.S.A.
[1] *Present address: New York Department Environmental Conservation, Stamford, NY 12167, U.S.A.*

Keywords: Environmental effects, Food, Growth, Morphometric Comparisons, Status, Threats to survival

Synopsis

Scaphirhynchus albus and *S. platorynchus* were studied in Missouri during 1978–1979 to assess their distribution and abundance, to obtain information on their life histories, and to identify existing or potential threats to their survival. *S. platorynchus* was collected in substantial numbers (4355 specimens) at all 12 sampling stations in the Missouri and Mississippi rivers, while only 11 *S. albus* were captured from 6 stations. Twelve specimens identified in the field as hybrids between the two species were captured from 4 stations. Morphometric and meristic comparisons of presumed hybrids with the parent species, using cluster and principal components analyses, demonstrated intermediacy of most specimens identified in the field as hybrids. Aquatic insects comprised most of the diet of *S. platorynchus* and *S. albus*, but *S. albus* and the hybrids had consumed considerable quantities of fish. *S. albus* grew more rapidly than *S. platorynchus*, while the growth of hybrids was intermediate. Hybridization appears to be a recent phenomenon, resulting from man-caused changes in the big-river environment. Hybridization may be a threat to survival of *S. albus* in the study streams.

Introduction

The river sturgeons, *Scaphirhynchus albus* and *S. platorynchus*, are endemic to large rivers of the central United States, primarily in the Mississippi Basin (Bailey & Cross 1954). *S. platorynchus* occurs widely in the Mississippi River and its major tributaries, and is sympatric with *S. albus* in the middle and lower Mississippi, Missouri, and Yellowstone rivers.

S. platorynchus was abundant and supported a substantial commercial fishery in the early part of this century (Carlander 1954). Today, this fishery is reduced, but is still important locally (Helms 1974); *S. platorynchus* is still rather common over much of its range. *S. albus* was not recognized as a species until 1905 (Forbes & Richardson 1905), and has not been distinguished in commercial fishery reports. Perhaps it has never been especially abundant, and it is one of the lesser known freshwater fishes in North America (Kallemeyn 1983).

In Missouri, the most recent authenticated capture of *S. albus* prior to this investigation was in 1948 (Fisher 1962), and this species has been classified as endangered (Nordstrom et al. 1977). This study of *S. albus* was undertaken to better define its distribution and abundance within the state, to obtain information on its life history, and to identify existing or potential threats to its survival. Comparative information was obtained on *S. platorynchus* because of its close relationship to *S. albus*, and because of the opportunity for study afforded by the planned investigation of its more seriously threatened relative.

An unexpected result of this investigation was the capture of apparent hybrids between the two species. Although many hybrid combinations are known between species of sturgeons (Acipenseridae), none have been previously reported in the genus *Scaphirhynchus* (Schwartz 1972, 1981). In this report we document our findings concerning possible hybridization in this genus, and evaluate these findings with respect to the status and prospects for survival of *S. albus* in the lower Missouri and Mississippi rivers.

Materials and methods

Sturgeons were collected from 12 stations in the Missouri and Mississippi rivers, in and adjoining the state of Missouri (Table 1). Most sturgeons were obtained with trotlines, trammel nets, and dip nets. Trotlines were baited with worms and fished on the bottom over sandbars. Trammel nets were weighted and set to fish near the bottom behind wing dikes, or were drifted near the bottom in the main channel. At the Chain of Rocks station, sturgeons entrapped in the forebay of the St. Louis City water intake were collected by dip netting. These gear types were fished a total of 95 days during spring and fall of 1978 and 1979.

The fork length (FL) of all captured sturgeons were recorded, and most specimens were released. A representative series of 5-15 individuals from each station were retained for internal examination. From each of these fish, the gonads were examined to determine sex and reproductive condition, the stomach was preserved for food habit studies, and the first pectoral-fin ray was kept for aging. Some specimens were preserved whole for meristic and morphometric comparisons, and tissue samples from selected specimens were frozen for electrophoretic analysis.

In the laboratory, stomachs were transferred from the field preservative (10% formalin) to 70% ethanol. The volume of the stomach contents was determined, food items were sorted and identified to the lowest possible taxonomic level, and the percentage each comprised of the total volume was estimated visually. The quantities of material too far digested for identification, as well as earth-

Table 1. Numbers of *S. platorynchus*, *S. albus*, and presumed hybrids obtained at 12 stations on the Missouri and Mississippi rivers.

Station[1]	*S. platorynchus*	*S. albus*	Hybrids	Total
Missouri River				
Brownsville (RM 534)	481	2	0	483
St. Joseph (RM 461)	65	1	0	66
Kansas City (RM 360)	347	1	2	350
Brunswick (RM 258)	148	0	0	148
Easley (RM 172)	608	1	0	609
St. Louis (RM 16)	148	0	2	150
Mississippi River				
Canton (RM 341)	46	0	0	46
Saverton (RM 302)	331	0	0	331
Chain of Rocks (RM 188)	1800	0	7	1807
Ste. Genevieve (RM 118)	97	1	0	98
Cairo (RM 952)	196	5	1	202
Caruthersville (RM 852)	65	0	0	65
Totals	4332	11	12	4355

[1] Station location designated by nearest city and river mile (RM) of station midpoint, taken from U.S. Army Corps of Engineers navigation charts.

worms (from specimens caught on baited hooks) were excluded from food habit comparisons.

Pectoral-fin rays were sectioned, and the sections were cleared in xylene and emersed in glycerine for age determinations. Annuli were counted toward the anterior apex of the fin-ray section to minimize errors in counts (Sokolov & Akimova 1976), and ages were reported for fish size at time of capture.

To document possible hybridization and confirm field identifications, meristic and morphometric comparisons of preserved specimens were made, using techniques and characters proposed by Bailey & Cross (1954). Four fin-ray counts and 10 measurements were made on each specimen (see Table 2 for a list of characters). Measurements were converted to thousandths of standard length (SL). The development of scutes on the belly were quantified by establishing four character states and ranking them on a scale of 1–4. These character states were: (1) belly with a mosaic of well developed scutes; (2) belly as in (1), but with a naked strip anteriorly at the midline; (3) belly with a few widely scattered scutes; and (4) belly naked except for a few rudimentary scutes posteriorly.

Cluster analysis and principal components analysis were performed, using computer programs provided by SAS (Ray 1982). All characters (except standard length) listed in Table 2 were used as variables in each analysis unless otherwise noted. For all analyses, field identifications were used in designating specimens as *S. platorynchus*, *S. albus*, or hybrids.

Frozen tissue samples from 7 *S. platorynchus*, 10 *S. albus*, and 6 presumed hybrids were sent for electrophoretic analysis to the population genetics laboratory at the University of Montana. The findings of this analysis are reported in detail elsewhere (Phelps & Allendorf 1983).

Table 2. Univariate statistics and average coefficient of variation (C.V.) for two species of *Scaphirhynchus* and hybrids between them. Group assignments are based on field identifications. Morphometric characters are expressed as thousandths of standard length. $\bar{x}$ is mean, $s_{\bar{x}}$ standard error of the mean, and N is sample size.

Character	*S. platorynchus* N = 10		Hybrids N = 12		*S. albus* N = 8	
	$\bar{x} \pm 2s_{\bar{x}}$	C.V.	$\bar{x} \pm 2s_{\bar{x}}$	C.V.	$\bar{x} \pm 2s_{\bar{x}}$	C.V.
Meristic						
Dorsal-fin rays	32.4 ± 1.6	7.7	37.6 ± 1.5	6.8	38.4 ± 1.3	5.0
Anal-fin rays	20.3 ± 0.8	6.6	23.5 ± 0.7	5.3	24.5 ± 0.9	5.3
Pectoral-fin rays	40.9 ± 2.8	10.8	45.8 ± 2.1	8.1	46.0 ± 3.1	9.6
Pelvic-fin rays	26.8 ± 1.6	9.8	27.9 ± 1.1	6.6	30.5 ± 0.9	4.3
Belly scutellation	1.6 ± 0.6	60.4	3.0 ± 0.4	24.6	3.9 ± 0.2	9.1
Morphometric						
Head length	262.3 ± 6.9	4.1	281.1 ± 10.4	6.4	300.1 ± 8.8	4.1
Rostral length	179.0 ± 8.7	7.7	197.2 ± 6.2	5.5	205.4 ± 6.2	4.2
Orbital length	9.7 ± 2.2	35.4	11.6 ± 0.9	13.5	9.6 ± 1.9	28.8
Mouth width	75.0 ± 2.3	4.8	85.0 ± 3.4	6.9	93.1 ± 6.6	10.1
Snout to outer barbel	100.2 ± 3.4	5.3	119.0 ± 6.8	9.9	133.3 ± 5.1	5.4
Mouth to inner barbel	60.6 ± 3.3	8.5	59.2 ± 2.8	8.3	55.9 ± 2.5	6.4
Outer barbel length	84.6 ± 7.5	14.0	103.6 ± 7.7	12.9	114.6 ± 19.9	24.6
Inner barbel length	62.5 ± 3.9	10.0	65.1 ± 2.3	6.2	56.6 ± 7.7	19.2
Tenth lateral plate	46.7 ± 3.3	11.3	39.9 ± 1.1	4.7	34.9 ± 2.4	9.6
Average C.V.		14.0		9.0		10.4
Standard length (mm)	580.0		588.0		622.6	

Results

Identification and documentation of hybrids

We captured 4355 river sturgeons in this study. The field identifications of these specimens were: 4332 (99.5%) *S. platorynchus*, 11 (0.2%) *S. albus*, and 12 (0.3%) hybrids between the species (Table 1). Sturgeons were identified as hybrids because of intermediacy or inconsistency in the expression of certain characters (barbel placement and length, rostral shape and length, belly scutellation, and coloration) that are readily observed under field conditions. Specimens of the presumed hybrids and the parent species were subjected to morphometric, meristic, and biochemical comparisons to substantiate or refute the field identification.

Thirty river sturgeons ranging in standard length from 447 to 816 mm were used in comparisons. The presumptive identifications of these specimens were: 10 *S. platorynchus*, 8 *S. albus*, and 12 hybrids. In all characters except orbital length and inner barbel length, the mean for presumed hybrids was intermediate between the means for *S. platorynchus* and *S. albus* (Table 2). However, for most of these characters the hybrid mean was closer to that of *S. albus* than to that of *S. platorynchus*. The means for orbital length and inner barbel length were greater in the hybrids than in the two species. The presumed hybrids did not exhibit greater variability in most characters than did the species. In 9 of 14 characters and in the mean coefficient, *S. platorynchus* exhibited the highest coefficient of variation of the three sturgeon groups.

Cluster analysis, using all 14 characters listed in Table 2, produced a grouping of the specimens that is quite consistent with the field identifications (Fig. 1). When three clusters are assumed, 9 (90%) of 10 *S. platorynchus* were allocated to one cluster, 9 (75%) of 12 presumed hybrids were allocated to a second cluster, and 6 (75%) of 8 *S. albus* were allocated to a third cluster. Of the total sample of 30 specimens, 6 (20%) were allocated to a different group by cluster analysis than by the field identification.

Principal components were calculated for all

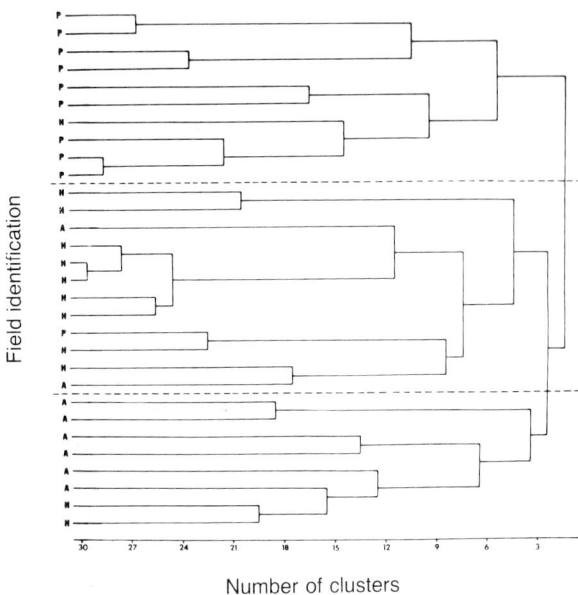

Fig. 1. Dendrogram for 30 *Scaphirhynchus*, produced by application of the SAS cluster procedure to 5 meristic and 9 morphometric characters. The field identifications of specimens were: P = *S. platorynchus*, A = *S. albus* and H = hybrid.

characters, and the scores were plotted for the first two principal components (Fig. 2). This resulted in a complete separation of specimens identified as *S. platorynchus* from those identified as *S. albus*, with most of the presumed hybrids isolated between them. However, two specimens identified as hybrids are plotted within the *S. albus* group, while

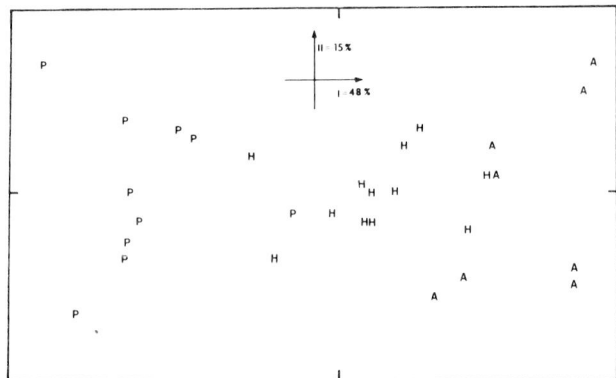

Fig. 2. Relative similarity of 30 *Scaphirhynchus*, as shown by projections on the first two principal components of the character correlation matrix for 5 meristic and 9 morphometric characters. The field identifications of specimens were: P = *S. platorynchus*, A = *S. albus*, and H = hybrid.

one *S. platorynchus* is within the hybrid group. One presumed hybrid was well separated from the hybrid group and near the *S. platorynchus* group.

The variation shown in Figure 2 accounts for 48 + 15 = 63% of the total variance in the data matrix. All characters except orbital length and inner barbel length contributed significantly (P<0.05) to principal component 1 (Table 3). Five characters (pelvic-fin rays, orbital length, mouth to inner barbel, outer barbel length and inner barbel length) contributed significantly to principal component 2. Three of these characters contributed significantly to both components.

To provide some perspective for interpreting these results, we calculated principal components for 15 specimens longer than 250 mm SL listed by Bailey & Cross (1954). Since only morphometric characters were available for their fish, this required recomputing the principal components for our fish, using only morphometric characters.

When scores of the first two principal components are plotted as before, greater overlap is evident between the three sturgeon groups (Fig. 3) than was the case when all characters were used (Fig. 2). *S. platorynchus* and *S. albus* are still well separated, but the presumed hybrids more broadly

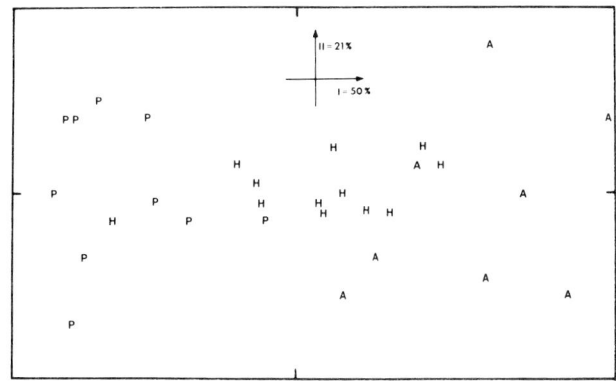

Fig. 3. Relative similarity of 30 *Scaphirhynchus*, as shown by projections on the first two principal components of the character correlation matrix for 9 morphometric characters. The field identifications of specimens were: P = *S. platorynchus*, A = *S. albus*, and H = hybrid.

overlap the plots for the two species.

When scores are plotted for the specimens of Bailey & Cross (1954), two widely separate groups, corresponding to *S. platorynchus* and *S. albus*, are evident (Fig. 4). The plot for the Bailey & Cross (1954) specimens appears comparable to the plot for specimens identified as *S. platorynchus* and *S albus* in our study. However, 9 of their 15 specimens were smaller than any used in our analysis,

Table 3. Correlation coefficients for the first two principal components of *Scaphirynchus* from the present study and from specimens examined by Bailey & Cross (1954). Significant correlations (P<0.05) are indicated by an asterisk.

Character	Present study				Bailey & Cross (1954)	
	All characters		Morphometric characters		Morphometric characters	
	Component 1	Component 2	Component 1	Component 2	Component 1	Component 2
Dorsal-fin rays	.79*	.18				
Anal-fin rays	.84*	−.03				
Pectoral-fin rays	.51*	−.03				
Pelvic-fin rays	.66*	−.42*				
Belly scutellation	.80*	−.16				
Head length	.82*	.02	.93*	−.05	.93*	−.04
Rostral length	.76*	−.25	.81*	−.27	.84*	−.40
Orbital length	−.06	−.68*	−.07	−.65*	−.48	−.73*
Mouth width	.85*	.31	.89*	.23	.85*	.18
Snout to outer barbel	.89*	−.06	.94*	−.15	.98*	−.06
Mouth to inner barbel	−.40*	.55*	−.30	.55*	−.84*	−.16
Outer barbel length	.68*	.64*	.70	.62	.41	.76*
Inner barbel length	−.13	.67*	−.18	.80*	−.89*	.26
Tenth lateral plate	−.78*	−.11	−.84*	−.04	−.87*	.33

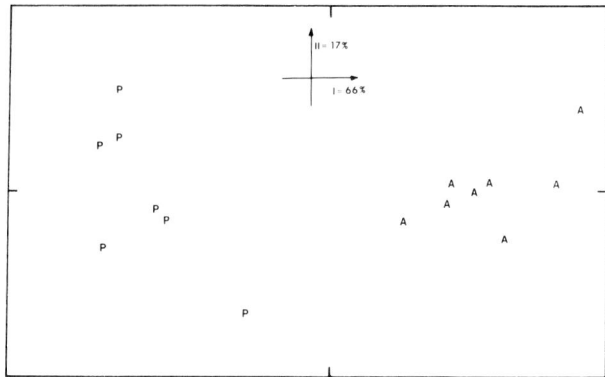

Fig. 4. Relative similarity of 15 *Scaphirhynchus*, as shown by projections on the first two principal components of the character correlation matrix for 9 morphometric characters. The identifications of specimens were: P = *S. platorynchus* and A = *S. albus*. The data for this analysis came from Bailey & Cross (1954).

and their smallest specimens exhibited the most intermediate principal component scores. Perhaps the separation between the species plots for their data would have been even greater if larger specimens had been available. There is general agreement between their data and our data in the contribution of various morphometric characters to component 1 and component 2 (Table 3). The major exceptions appear to involve characters relating to barbel position and length.

Electrophoretic analysis revealed that *S. platorynchus* and *S. albus* were indistinguishable at all 37 loci examined (Phelps & Allendorf 1983). Also, at three polymorphic loci, *S. albus* and *S. platorynchus*, and geographically distant populations of the latter species, exhibited no statistically significant differences in morph frequency. Therefore, electrophoresis provided no supportive evidence for or against the identification of some specimens as hybrids. The similarity at such a large number of loci suggests a close genetic relationship between these two species. This result is surprising, considering the many phenotypic differences between them.

Distribution, relative abundance, and habitat

S. platorynchus occurs throughout the Missouri and Mississippi rivers in Missouri, and was captured in substantial numbers at all 12 stations (Table 1). This species comprised 73% of all fishes in our collections. However, our sampling techniques and gear were intentionally selective for sturgeons.

S. albus was recorded at 6 of 12 stations, and 11 specimens were captured. The largest number of specimens (5) were from the Mississippi River at Cairo, where this species comprised 2.5% of the river sturgeons in our collections. The proportions of *S. albus* in collections from other stations were: Brownsville – 0.4%, St. Joseph – 1.5%, Kansas City – 0.3%, Easley – 0.2%, and Ste. Genevieve – 1.0%. No *S. albus* were captured from the Mississippi River upstream from the mouth of the Missouri River.

Twelve sturgeons identified as hybrids were collected from four stations. The most specimens (7) came from Chain of Rocks on the Mississippi River. At that station they comprised 0.4% of the river sturgeons captured. The proportions of hybrids at other stations were: Kansas City – 0.6%, St. Louis – 1.3%, and Cairo – 0.5%.

Both *S. platorynchus* and *S. albus* were found in the main channels of the river, along sandbars at the inside of river bends and behind wing dikes with deeply scoured trenches. *S. albus* was generally taken in gear-sets (usually trotlines) that contained *S. platorynchus*, but four of eleven specimens were caught in areas with swift current where *S. platorynchus* was less numerous. Sturgeons identified as hybrids appeared to be more closely associated with *S. platorynchus* than with *S. albus*. On the average, each gear-set that caught a hybrid contained 14 *S. platorynchus*, compared with an average of two *S. platorynchus* for gear-sets that caught *S. albus*.

Food, growth, and sex ratio

Aquatic invertebrates (principally the immature stages of insects) comprised most of the diet of river sturgeons captured in this study, but with a greater proportion of fish (mostly cyprinids) in the diet of *S. albus* and presumed hybrids than in *S. platorynchus* (Table 4). These differences in the consumption of fish were evident in both volume (T_{P-A}

− 2.5, df = 8) and frequency of occurrence (X^2_{P-A} = 36.4, df = 1; X^2_{P-H} = 5.7, df = 1). Coker (1930) and Cross (1967) also reported a high incidence of fish in the diet of *S. albus*, while other investigators (Held 1969, Walberg et al. 1971, Helms 1974, Modde & Schmulbach 1977, Durkee et al. 1979) reported a low incidence of fish (less than 2% by volume) in the diet of the *S. platorynchus*. Sand occurred frequently in the stomachs examined (frequency of occurrence 24.6% for all specimens combined) and was probably consumed incidentally along with food items. The occurrence of plant material in the stomachs may also have been incidental.

The length of *S. albus* was significantly greater than that of *S. platorynchus* (T = 4.77, df = 6) for each age group in which comparable data was available, while the hybrids were generally intermediate in length (Fig. 5). Fogle (1963) also reported more rapid growth in *S. albus* than in *S. platorynchus*.

Females predominated over males in our samples of *S. platorynchus* and *S. albus*, with males comprising 27% and 33%, respectively of all the specimens examined. Three percent of the *S. platorynchus* were hermaphrodites. All 12 of the presumed hybrids were females.

Discussion

These analyses show that *S. albus* and *S. plato-*

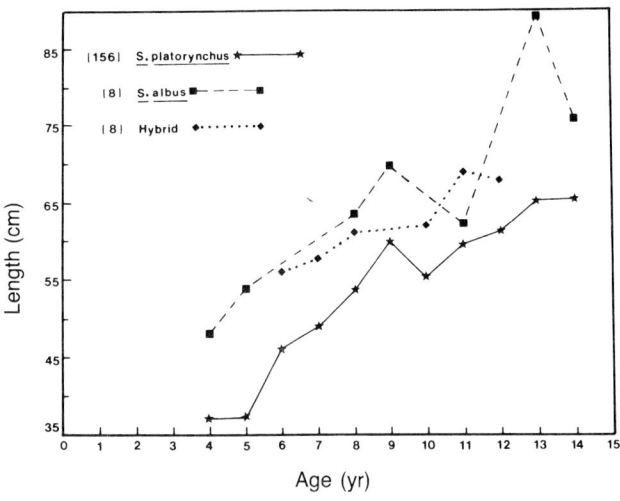

Fig. 5. Average length (FL) at each age for 172 *Scaphirhynchus* from the Missouri and Mississippi rivers.

rynchus hybridize. Specimens tentatively identified as hybrids in the field were intermediate in several meristic and morphometric characters. The hybrids appeared to be more like *S. platorynchus* in habitat selection, but their diet resembled that of *S. albus* in containing a substantial quantity of fish. The growth rate of hybrids was intermediate between that of the parent species. All of the hybrids were females, possibly indicating an unbalanced sex ratio, as has been reported in some other fish hybrids (Hubbs 1955).

A principal components analysis of morphometric data presented by Bailey & Cross (1954) demonstrated that their sample included two mor-

Table 4. Composition by volume and frequency of occurrence of food categories in the diet of *S. platorynchus* (N = 231), *S. albus* (N = 9) and presumed hybrids (N = 9).

Food category	Composition by volume (%)			Frequency of occurrence (%)		
	S. platorynchus	Hybrids	*S. albus*	*S. platorynchus*	Hybrids	*S. albus*
Ephemeroptera	17.3	16.5	6.0	55.1	88.9	44.4
Odonata	7.3	5.3	8.6	18.4	11.1	33.3
Plecoptera	3.4	2.2	0.0	16.7	22.2	0.0
Trichoptera	34.0	4.8	37.8	74.8	44.4	55.6
Diptera	19.1	4.9	3.0	81.2	77.8	44.4
Other insects	11.8	2.8	2.9	44.9	22.2	33.3
Other invertebrates	2.0	4.6	1.3	15.8	66.7	33.3
Fish	1.6	31.3	37.7	4.3	22.2	55.6
Plant material	3.4	27.6	2.6	39.7	22.2	55.6

phologically distinct types of *Scaphirhynchus*, while a comparable analysis of our specimens demonstrated that they were not similarly separable into two types. The failure of Bailey & Cross (1954) and other workers to report hybrids between the species of *Scaphirhynchus* may indicate that hybridization is a recent phenomenon, resulting from a fundamental change in the behavioral and ecological relationships between these species. These changes may be related to man-induced reductions in habitat diversity and measureable changes in environmental parameters such as turbidity, flow regimens and substrate types. Projects by the U.S. Army Corps of Engineers designed to deepen and stabilize the lower Missouri River have reduced the water surface area by 50% and largely eliminated the numerous islands and side channels that were formerly present (Funk & Robinson 1974). Six large mainstem reservoirs constructed on the upper river have modified the natural seasonal flood patterns and resulted in measurable reductions in turbidity all the way to the river mouth (Neel et al. 1963, Whitley & Campbell 1973). Similar changes are evident in the Mississippi River downstream from the mouth of the Missouri. Schmulbach (1974) in discussing suspected hybridization between *Stizostedion canadense* and *S. vitreum* in the Missouri River, South Dakota stated that hybridization between species seems to be limited to places where man or nature has 'hybridized the habitat'. He concluded that the Missouri River is such a hybridized habitat.

Presumed hybrids were as prevalent in our samples as *S. albus*, suggesting that hybridization between the species of *Scaphirphynchus* may occur frequently. This hybridization could present a threat to survival of *S. albus*, through genetic swamping if the hybrids are fertile, and through competition for a limited habitat. Studies are needed to determine fertility of the hybrids and the extent and consequences of backcrossing.

Acknowledgements

Funds for this study were provided by the Office of Endangered Species, U.S. Fish and Wildlife Service (Endangered Species Project SE-1-6) and the Missouri Department of Conservation. We thank Mike Petersen for assistance in the field and laboratory, Stevan R. Phelps for doing the electrophoretic analysis, Frank B. Cross for loaning specimens, and Reeve M. Bailey for providing unpublished data. Joe G. Dillard, Thomas R. Russell, and Walter T. Keller reviewed early drafts of the manuscript.

References cited

Bailey, R.M. & F.B. Cross. 1954. River sturgeons of the American genus *Scaphirhynchus:* characters, distribution, and synonomy. Pap. Mich. Acad. Sci., Arts, Let. 39: 169–208.

Carlander, H.B. 1954. A history of fish and fishing in the Upper Mississippi River. Spec. Publ., Upper Miss. R. Cons. Comm. 96 pp.

Coker, R.E. 1930. Studies of common fishes of the Mississippi River at Keokuk. U.S. Bur. Fish. Bull. 45: 141–225.

Cross, F.B. 1967. Handbook of fishes of Kansas. Mus. Nat. Hist. Univ. Kans., Misc. Publ. 45: 1–357.

Durkee, P., B. Paulson & R. Bellig. 1979. Shovelnose sturgeon (*Scaphirhynchus platorynchus*) in the Minnesota River. Minn. Acad. Sci. 45(2): 18–20.

Fisher, H.J. 1962. Some fishes of the lower Missouri River. Amer. Midl. Nat. 68: 424–429.

Fogle, N.E. 1963. Report of fisheries investigations during the fifth year of impoundment of Oahe Reservoir, South Dakota 1962. South Dakota Dept. Game Fish and Parks. D-J Proj. F-1-R-12, Job 10-12. 35 pp.

Forbes, S.A. & R.E. Richardson. 1905. On a new shovelnose sturgeon from the Mississippi River. Bull. Ill. State Lab. Nat. Hist. 7: 37–44.

Funk, J.L. & J.W. Robinson. 1974. Changes in the channel of the lower Missouri River and effects on fish and wildlife. Mo. Dept. Cons. Aquatic Ser. 11. 52 pp.

Held, J.W. 1969. Some early summer foods of the shovelnose sturgeon in the Missouri River. Trans. Amer. Fish. Soc. 98: 514–517.

Helms, D.R. 1974. Shovelnose sturgeon, *Scaphirhynchus platorynchus* (Rafinesque) in the navigational impoundments of the Upper Mississippi River. Iowa Cons. Comm. Tech. Ser. 74-3. 68 pp.

Hubbs, C.L. 1955. Hybridization between fish species in nature. Syst. Zool. 4: 1–20.

Kallemeyn, L. 1983. Status of the pallid sturgeon. Fisheries 8: 3–9.

Modde, T. & J.C. Schmulbach. 1977. Food and feeding behavior of the shovelnose, sturgeon, *Scaphirhynchus platorynchus*, in the unchannelized Missouri River, South Dakota. Trans. Amer. Fish. Soc. 106: 602–608.

Neel, J.K., H.P. Nicholson & A. Hirsch. 1963. Main stem reservoir effects on water quality in the central Missouri

River. U.S. Dept. Health, Education, and Welfare Publ., Kansas City. 112 pp.

Nordstrom, G.R., W.L. Pflieger, K.C. Sadler & W.H. Lewis. 1977. Rare and endangered species of Missouri. Mo. Dept. Cons., Jefferson City. 129 pp.

Phelps, S.R. & F. Allendorf. 1983. Genetic identity of pallid and shovelnose sturgeon (*Scaphirhynchus albus* and *S. platorynchus*). Copeia 1983: 696–700.

Ray, A.R. 1982. SAS user's guide: statistics, 1982 edition. SAS Institute, Cary. 584 pp.

Schmulbach, J.C. 1974. An ecological study of the Missouri River prior to channelization. S. Dak. Water Resources Proj. P-024, Univ. S. Dak. Completion Rep. 34 pp.

Schwartz, F.J. 1972. World literature to fish hybrids with an analysis by family, species, and hybrid. Publ. Gulf Coast Res. Lab. Mus. 3: 1–328.

Schwartz, F.J. 1981. World literature to fish hybrids with an analysis by family, species, and hybrid: Supplement 1. NOAA Tech. Rep. NMFS SSF 750: 1–507.

Sokolov, L.I. & N.V. Akimova. 1976. Age determination of the Lena River sturgeon *Acipenser baeri*. J. Ichthyol. 16: 773–778.

Walburg, C.H., G.L. Kaiser & P.L. Hudson. 1971. Lewis and Clark Lake tailwater biota and some relations of the tailwater and reservoir fish population. pp. 449–467. *In:* G.E. Hall (ed.) Reservoir Fisheries and Limnology, Amer. Fish. Soc. Spec. Publ. 8, Bethesda.

Whitley, J.R. & R.S. Campbell. 1973. Some aspects of water quality and biology of the Missouri River. Trans. Mo. Acad. Sci. 7–8: 60–72.

Received 15.5.1984   Accepted 15.1.1985

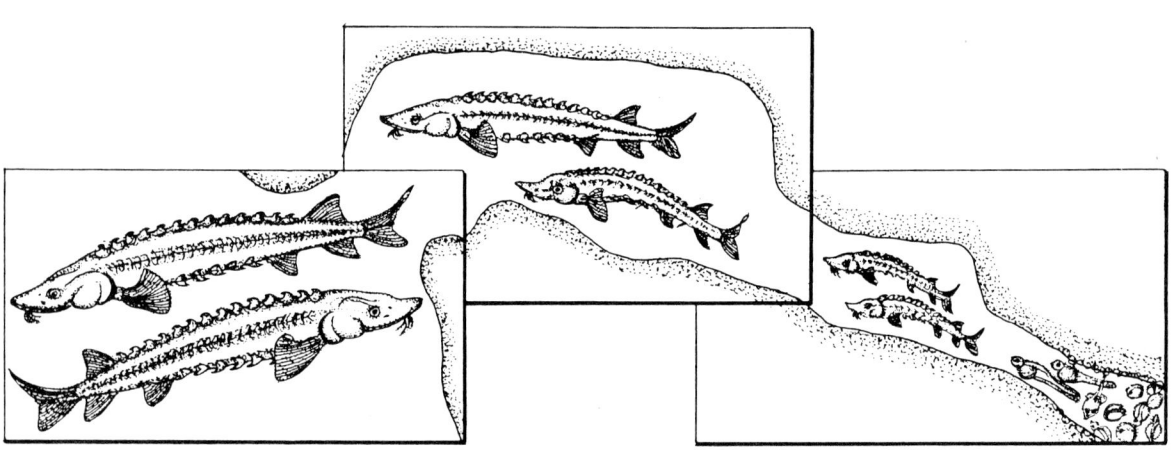

Sturgeon spawning occurs primarily in large rivers. The embryos hatch in 7 to 15 days depending on temperature. The young fish are found in the river and estuary until late fall at which time they migrate to deeper waters.

The fishery, biology, and management of Atlantic sturgeon, *Acipenser oxyrhynchus*, in North America

Theodore I.J. Smith
South Carolina Wildlife and Marine Resources Department, Marine Resources Research Institute, P.O. Box 12559, Charleston, SC 29412, U.S.A.

Keywords: Acipenseridae, Life history, Ecology, Harvest, Culture

Synopsis

The Atlantic sturgeon supported major fisheries along the entire Atlantic coast of North America. These fisheries peaked about 1890 and then suffered almost total collapse by 1905. The Atlantic sturgeon is anadromous and highly susceptible to capture during spawning migrations. Further, this species biological characteristics makes it very vulnerable to man-induced changes in natural habitat and slow to recover. Atlantic sturgeon mature at an advanced age (7–27 year for females, depending on latitude), exhibit a long interspawning period (2–5 year), and require suitable riverine, estuarine, and coastal environments for successful completion of their life cycle. Today, only remnant stocks exist in areas of former abundance. Management regulations vary considerably from state to state and range from full protection to no protection. Biological data are needed to: identify and characterize specific spawning and nursery areas; delineate migratory patterns and recruitment to various stocks; establish stock abundance; and, assess effects of various management strategies. In order to protect remaining stocks, the imposition of a total harvesting moratorium is recommended.

Introduction

During colonial days Atlantic sturgeon, *Acipenser oxyrhynchus*, were abundant and served as an item of commerce. Evidence of this is recorded in the early literature dealing with the natural history and settlement of the Atlantic coast of North America. For example, in 1634 William Wood published a book titled 'New England's Prospect' in which he commented: 'The sturgeons be all over the country, but the best catching of them be on the Shoales of Cape Codde and in the Rivers of Merrimacke, where much is taken, pickled and brought for England some of these be twelve, fourteen, eighteen foote long'. Later, in 1675, Gent commented 'This fish is here in great plenty and in some rivers so numerous that it is hazaradous for canoes and the like small vessels to past too and again. (...)'. Early coastal dwellers were very knowledgeable in the migratory behavior of Atlantic sturgeon and in methods of capture, processing, and shipping. Unfortunately, because of man's earlier activities, todays' coastal residents rarely encounter this species and for the most part are totally unaware of its existence.

The Atlantic sturgeon is represented by two subspecies. The northern subspecies, *Acipenser oxyrhynchus oxyrhynchus*, is more abundant and more widely distributed, ranging from Hamilton Inlet on the Atlantic coast of Labrador (Bachus 1951), throughout the Gulf of St. Lawrence and in the St. Lawrence River, in Nova Scotia waters and in the

St. John River, New Brunswick, as well as at the head of the Bay of Fundy (Murawski & Pacheco 1977). Their range continues along the entire Atlantic coast to the St. Johns River in eastern Florida with an occasional specimen reported from Bermuda (Bigelow & Schroeder 1953, Vladykov & Greeley 1963). In contrast, the southern subspecies, *A. o. desotoi,* occupies a much more restricted range and is limited to the Gulf of Mexico and northern coast of South America (Huff 1975). Chief differences between subspecies are shape of scutes, length of pectoral fins, relative head length (Vladykov & Greeley 1963) and most significantly, length of spleen relative to fork length (16–19% in Gulf sturgeon, 3–9% in Atlantic sturgeon (Wooley & Crateau 1982, Wooley 1985).

During recent years, there has been a renewed interest and awareness in Atlantic coast sturgeons for several reasons. First, both the Atlantic and shortnose sturgeon, *A. brevirostrum* (listed as an endangered species in the U.S. – Miller 1972), inhabit riverine and coastal environments which are often selected as sites for construction of electric generating stations and other industrial activities. Consequently, information on their occurrence and potential response to planned activities must be addressed in environmental impact statements and construction permit applications. Second, fishery managers have become aware that the sturgeon fisheries were one of the major fisheries along the entire Atlantic coast and that now most stocks are depleted and facing local extinction in some areas. Third, substantial development activities are being focused on aquaculture and sturgeons are being examined as possible candidates for rearing in controlled culture systems, net pens, and/or in ocean ranching operations. For the most part, research and data gathering support has come from the U.S. Fish and Wildlife Service, the National Marine Fisheries Service, public utilities, and by some state natural resource agencies. Such support has broadened our knowledge of this species and resulted in the preparation of several recent review manuscripts (see Murawski & Pacheco 1977, Hoff 1980, Rulifson & Huish 1982, Van Den Avyle 1983). Additionally, Knox & Dadswell (1980) have provided an annotated bibliography dealing with the reproduction and early life history of sturgeons with reference to culture activities in the U.S.S.R.

This manuscript provides a brief summary of the fisheries and biological data on this species and discusses management recommendations.

Fisheries

Landings, gear, and products

Historically, sturgeon fisheries occurred in all major coastal rivers along the Atlantic coast. Fishing was seasonal, coinciding primarily with spawning migrations of this anadromous species into and from coastal and riverine systems. In the U.S., major fisheries were established around 1880 in New York, New Jersey, Pennsylvania, Delaware, Maryland, Virginia, North Carolina, South Carolina, and Georgia. These fisheries peaked about 1890 when total recorded landings were about 3.3 million kg (Fig. 1). By the turn of the century, all U.S. fisheries had suffered substantial decreases in landings and most of the major fisheries exhibited almost total collapse. During the peak harvesting period of 1880–1890, New Jersey, Delaware, and Virginia were the prime producers of sturgeon based mostly on landings from the Delaware River and to a lesser extent from the Chesapeake Bay system. During 1978–1982, however, these states ranked only fifth, tenth, and sixth, respectively, with reported landings less than 1% of their former levels (Table 1). In recent years, South Carolina and North Carolina provided about 70% of the U.S. landings but in 1983 and 1984, South Carolina landings were severely depressed (Smith unpublished). Fishery landings for Atlantic sturgeon in Canada have also showed a drastic decline (Fig. 1) with the St. John River in New Brunswick providing about half of the total Canadian landings (Mike Dadswell, personal communication). Currently, landings of Atlantic sturgeon in North America are on the order of 100 to 130 000 kg with the U.S. providing about 80–90% of the landings.

The sturgeon fisheries employed a wide diversity of gear to catch the migrating sturgeon. The most popular gear was large floating or anchored gill

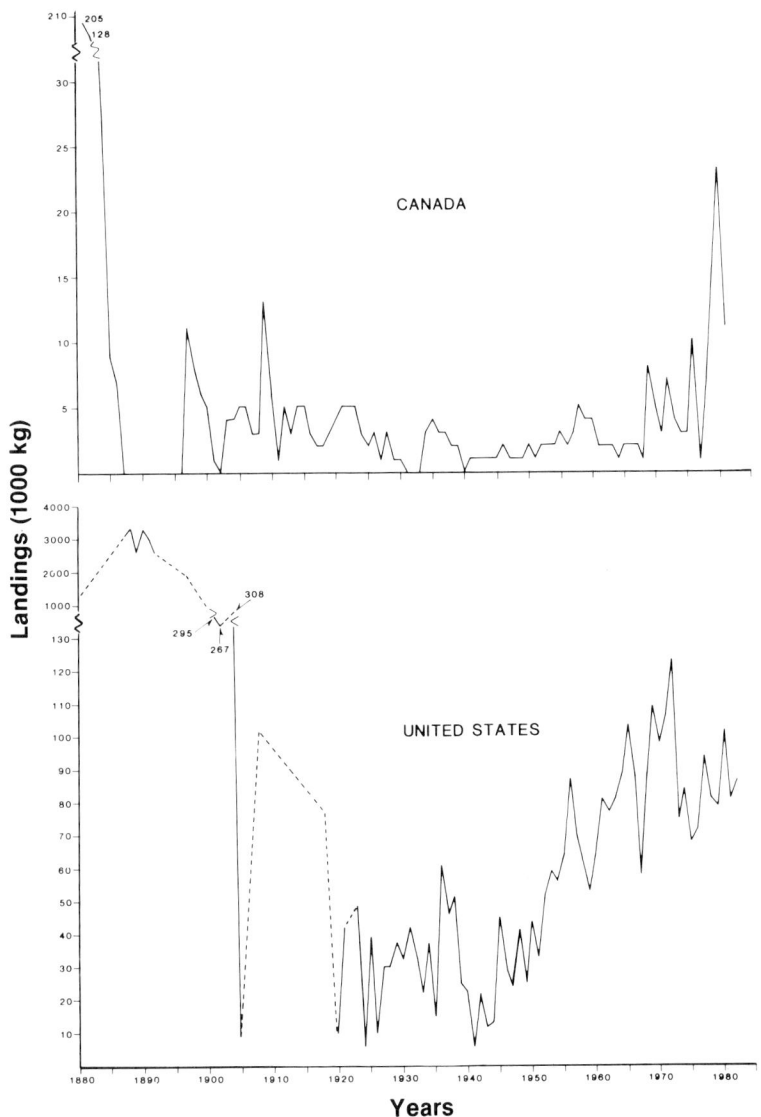

Fig. 1. Reported landings of Atlantic sturgeon from the United States and Canada (St. John River). Pre-1973 landings may include shortnose sturgeon. Landings from the St. John River represent about 50% of total Canadian landings (M. Dadswell, personal communication).

nets having a stretch mesh of about 33–41 cm, a depth of 4–8 m, and a length up to about 460 m. Other types of fishing gear included pound nets, trammel nets, weirs, stake row nets, trawls, and harpoons. Today, gill nets are still popular and are fished in ocean and river areas from small boats (Smith et al. 1984). A variety of products were made from sturgeon: the flesh was pickled, smoked, and served fresh; the eggs or roe were salted and sold as 'caviar'; oil was extracted by boiling the head, skin, and backbone; and, isinglass was obtained from the swim bladder and used as a clarifying agent and adhesive. Presently, the flesh and caviar are the main products.

Reasons for decline

Sturgeon are highly susceptible to man's activities as they reach maturity at an advanced age, exhibit a protracted spawning periodicity, and require suita-

Table 1. Average reported landings of Atlantic sturgeon by major producing states in the U.S. during 1888–1892[1] and 1978–1982[2].

Rank	1888–1892		1978–1982	
	State	Landings (1,000 kg)	State	Landings (1,000 kg)
1	New Jersey	1,892	South Carolina	45
2	Delaware	706	North Carolina	15
3	Virginia	350	New York	9
4	South Carolina	114	Georgia	7
5	North Carolina	102	New Jersey	5
6	Georgia	71	Virginia	2
7	Pennsylvania	27	Maine	1
8	Maryland	27	Massachusetts	1
9	New York	15	Maryland	1
10	Florida	11	Delaware	1

[1] Based on Murawski & Pacheco (1977); may include shortnose sturgeon, *A. brevirostrum*
[2] National Marine Fishery Service data.

ble riverine, estuarine, and coastal habitats to complete their life cycle. As a consequence, it has been mans' activities which have been directly responsible for the great decline in Atlantic sturgeon populations. During the 19th century, construction of dams on spawning rivers, especially in the northeast, effectively blocked passage of mature fish and restricted the amount of suitable spawning habitat. Examples include the dam at the head of the Androscoggin River (1807) and the Kennebec River (1837) in Maine, the Enfields Rapids Dam on the Connecticut River, and the dam at Lawrence (1847) on the Merrimack in New Hampshire (Hoover 1938, Galligan 1960, Murawski & Pacheco 1977). Since 1870, placement of mill dams and water supply dams on Peedee, Wateree, Congaree, and Savannah rivers have also precluded sturgeon in South Carolina from historical spawning sites (Leland 1968). Along the middle Atlantic region, water pollution from industrial and domestic sources along coastal areas also impacted Atlantic sturgeon by affecting spawning and nursery grounds. Nevertheless, of all considerations, overfishing seems to be the most important factor responsible for the drastic reduction in most sturgeon fisheries over the relatively short period of 20 to 25 years. This conclusion was alluded to by Ryder (1890) in his initial assessment of the sturgeon fisheries and restated recently by Hoff (1980) in his review of the current status of Atlantic sturgeon stocks.

Biology

Spawning and early development

The Atlantic sturgeon is truly anadromous, spawning in fresh or brackish water (Dean 1893, Dees 1961) but attaining most of its growth in saltwater. Exact spawning locations are generally unknown for most stocks of sturgeon, however, in the Hudson River, Dovel (1978, 1979) reports that most of the spawning activity occurs in freshwater just in front of the saltwater wedge. Timing of spawning migrations into coastal and riverine systems are well documented and show a distinct pattern which seem to be chiefly temperature regulated. Mature adults begin to arrive in Florida, Georgia, and South Carolina in February, in the Delaware and Chesapeake Bay systems in April, and in the Gulf of Maine and Gulf of St. Lawrence systems in May-June (Table 2). In South Carolina, water temperature at time of arrival of the first fish to the ocean area around the Winyah Bay jetties ranged from 7–10° C during 1979–1982 (Smith et al. 1982). Atlantic sturgeon spawn in running water over bottoms containing rocks, rubble, and other hard objects

Table 2. Time of spawning migrations of Atlantic sturgeon in North America.

Area	Beginning of migration	Source
Suwannee River, Florida	February	Huff (1975)
St. Mary's River, Georgia	February–March	Vladykov & Greeley (1963)
Winyah Bay, South Carolina	February–March	Smith et al. (1982)
Chesapeake Bay	April	Smith (1907), Hildebrand & Schroeder (1928)
Delaware	late April	Borodin (1925)
Hudson River	April–May	Vladykov & Greeley (1963), Dovel (1978, 1979)
Gulf of Maine	May–June	Bigelow & Schroeder (1953)
St. Lawrence	May–July	Scott & Crossman (1973)

and in pools below waterfalls (Dees 1961, Vladykov & Greeley 1963, Leland 1968, Huff 1975). In Florida, Wooley & Crateau (1982) have tentatively identified limestone shoal areas in the Appalachicola River as probable spawning grounds for the southern subspecies. Spawning is often accompanied by leaping and splashing apparently to help discharge the ovulated eggs. In the Delaware River spawning occurs at temperatures of 13.3–17.8° C (Borodin 1925) while in South Carolina ripe fish migrating to spawning areas have been captured at temperatures up to 21–23° C; however, most ripe fish were captured at 13–19° C. In Florida, one ripe male was captured at 20.6° C while a recently hatched sturgeon (9.71 mm) was captured at 23.9° C (Wooley & Crateau 1982). Mature eggs vary from gray to brown to black in color and range in size from about 2.5 to 3.0 mm in diameter. The extruded demersal eggs are highly adhesive and attach to rocks, gravel, plants, roots, etc. in ribbons or small clusters within twenty minutes after being broadcasted. The cleavage pattern is modified holoblastic with descriptions of eggs and development provided by Ryder (1890), Dean (1893), Mansueti & Hardy (1967), and Jones et al. (1978). Incubation period ranges from 94 h at 20° C (Dean 1894) to 168 h at 17.8° C (Vladykov & Greeley 1963). During recent culture trials in South Carolina, Smith et al. (1981) reported incubation times of 132 to 140 h at temperatures of 17.5–18° C.

The ripe ovaries can comprise as much as 25% of the total fish weight and contain a large number of eggs. Fecundity of fish captured in the Delaware River was estimated to vary from 800000 to 2400000 (Ryder 1890) while in North Carolina the fecundity estimates ranged from 1000000 to 2500000 eggs (Smith 1907). In South Carolina, Smith et al. (1982) estimated that a 100 kg female would produce about 1600000 eggs based on the formula, fecundity = 233064 + 13307 (fish weight in kg).

Little information is available on the early life history. Under culture conditions, sturgeon hatched at a mean size of 7.8 mm total length and grew to a size of 17.7 cm by day 204 (Smith et al. 1981). These authors also reported that the darkly pigmented hatched embryos were active swimmers, frequently leaving the bottom and swimming throughout the water column. Nine to ten days after hatching, the yolksac was absorbed and larvae begin to exhibit a benthic behavior.

Feeding, growth, and migration of juveniles

Young sturgeon primarily inhabit freshwater regions of rivers and eat a wide variety of plant and animal materials. Like the adults, juveniles root along the bottom sucking in materials through their ventral, protractile mouth. In the process, mud and plant materials are taken in as well as sludge worms *Limnodrilus*, chironomid and mayfly larvae, isopods, amphipods, and small bivalve mollusks (Scott & Crossman 1973). Information on the growth of small juveniles from tagging studies is lacking, however, in the St. Lawrence River, large juveniles ranging in size from 70.6–84.7 cm FL (2.2 to 4.0 kg) were tagged, released, and recaptured. Estimated annual length increment was 6.3 to 14.4% (weight gain 28.8 to 46.0%) (Scott & Crossman 1973).

Information on the movement of juveniles within the estuarine and riverine systems as well as along the coast has been documented through a number of tagging studies and summarized by Holland & Yelverton (1973), Huff (1975), Murawski & Pacheco (1977), Hoff (1980), and Rulifson & Huish (1982). In general, juveniles remain within their natal system but move progressively seaward with time. During the fall and winter months juveniles tend to congregate and move to deeper holes or channels as the temperature decreases. There is also some movement of juveniles into saline areas with decreasing temperatures and some immature animals enter coastal waters. Some juveniles leave their natal systems and exhibit coastal migrations apparently with some of these migrants moving into non-natal systems. In the St. Lawrence River, 1948 fish were tagged and released during 1945–1949 by Vladykov & Greeley (1963). They reported that migration of recaptured fish showed the typical pattern of a regular movement of juveniles towards freshwater in the spring (May–June) and back to saltwater in the fall (September–November). In the Hudson River, some tagged juveniles which emmigrated typically showed a southerly pattern along the coast in the fall moving as far as North Carolina and a suspected movement northerly in the spring (Dovel 1979). Fish tagged in North Carolina also moved south during November to January and then north along the coast during late winter and early spring (Holland & Yelverton 1973).

Juveniles are capable of traveling some distance over a relatively short period of time. For example, a sturgeon tagged in North Carolina (weight 9.5 kg) travelled 645 km to Long Island in 65 days (9.9 km per day). In South Carolina, a juvenile (59 cm FL) tagged in the Edisto River, was recaptured in Pamlico Sound, N.C. (326 days at large, 595 km travelled) while another juvenile (67 cm FL) tagged in the ocean area around the Winyah Bay jetties was recaptured in Chesapeake Bay (80 days at large, 807 km travelled) (Smith et al. 1982). In general, juveniles remain within the riverine systems for periods from about 1–6 years before emmigrating along the coast and onto the continental shelf where they grow to maturity.

Maturity, feeding, and growth of adults

Age at maturity of Atlantic sturgeon is influenced by sex and latitude. Males mature earlier than females with fish in northern latitudes reaching maturity at a much older age than those in the southern latitudes. In South Carolina, males first spawn at an average age of 8.1 year (range 5–13 year), while females first spawn at 10.9 year (range 7–19 year). In the Hudson River, males mature at age 11–20 year and females at 20–30 year (Dovel 1979), while in the St. Lawrence, males mature at 22–34 year and females at 27–28 year (Scott & Crossman 1973). In South Carolina, spawning marks on the pectoral rays indicate that females spawn at intervals ranging from about 3–5 year while males exhibit an average of 1–5 year between spawnings. This relatively long time between spawning is common among sturgeon. In South Carolina and elsewhere, it is normal for mature fish to participate in spring spawning migrations during years in which they are not going to spawn. Additionally, there are reports of a fall run or migration which occurs from about September through December, depending on latitude. In most cases, this is considered to be a post-spawning down-river migration. However, in South Carolina an up-river migration of ripe fish in late August and September has also been documented (Smith et al. 1984).

Adult Atlantic sturgeon are omnivorous benthic feeders which take in large quantities of mud during feeding activities. During spawning migrations these large fish do not feed and thus there is little recreational fishing. However, a small recreational snag fishery does exist just below the Jim Woodruff Lock and Dam on the Applachicola River in Florida (Wooley Post & Crateau 1982). Post-spawning adults, which remain in freshwater, feed, primarily on gastropods and other benthic organisms. In saltwater, diet components include amphipods, isopods, shrimps, polychaete worms, molluscs and small fishes, especially sand lances *Ammodytes* sp. (Scott & Crossman 1973). In the Suwannee River, the stomachs of pre-spawning migrating adults were reported to contain large amounts of partially digested vegetable materials mixed with hard parts of crabs (presumably *Callinectes sapidus*) (Huff 1975).

The Atlantic sturgeon is a long lived species capable of obtaining an advanced age and a large size. Growth data indicates a wide variation in growth rates by latitude with the most rapid growth exhibited by fish captured in the southern latitudes. For example, fish 7 year of age averaged 110 and 138 cm FL in the Suwannee River (Florida) and Winyah Bay systems respectively, while those in the St. John River (New Brunswick) and the St. Lawrence were 97 and 58 cm FL, respectively. At age 15, fish were 155 and 182 cm FL in the Suwannee River and Winyah Bay systems respectively, as compared to 105 cm in the St. Lawrence (Table 3). Fish have been aged to 60 year (Magnin 1964) and reference to early fishing efforts indicate that fish in the range of 4.3 to 5.3 m were not too infrequently caught. Unfortunately, such specimens may now exist only in museums. During the 1978–1982 South Carolina fishing seasons, average size and age of the commercially landed fish were as follows: females – 188 cm FL (range 124–233 cm), 71.8 kg (range 19–141), and age 16.0 year (8–30); males – 158 cm (90–203 cm), 41.2 kg (14–84 kg), 13.3 year (5–21 year) (Smith et al. 1984).

The relationship between weight and length has been calculated for sturgeon from the Suwannee River by Huff (1975) during 1972–1973 for both spring and fall captured fish by sex. He showed

Table 3. Mean fork length (cm) and age (y) data[1] for Atlantic sturgeon from different geographical areas (modified from Murawski & Pacheco 1977).

Age	Suwannee R.[2] Florida	Winyah Bay S.C.	Hudson R.[3] N.Y.	Kennebeck Maine	St. John R.[2] New Brunswick	St. Lawrence[3]
1	35	44			36	20
2	51	50	27		47	25
3	64	55	41		53	31
4	75	61	54	71	63	37
5	83	109	59	93	63	44
6	106		57	92	96	51
7	110	138	63		97	58
8	118	143	73	94	101	66
9	123	149			109	76
10	127	152	165	126	108	79
11	132	162	166	124	101	86
12	147	163	208		112	92
13	136	170				96
14	158	182				101
15	155	182				105
16	135	186			130	
17	149	193		151		
18		196				
19		202				
20		198	146			141
21		195				
22		201	143			
23		203				
24		199				
25		211				
46						226
60						233

[1] Data from Greeley (1937), Magnin (1964), Huff (1975), Squires & Smith (1979), Smith et al. (1982), M. Dadswell, personal communication.
[2] Approximate values.
[3] Converted from TL to FL using FL = 0.867TL + 10 mm (Magnin 1962).

significant differences between spring and fall fish, primarily due to weight loss. For mixed samples from North Carolina, Holland & Yelverton (1973) calculated $W = 5.46 \times 10^6 \, FL^{3.10}$ while $W = 1.14 \times 10^6 \, TL^{3.18}$ was derived for fish from the St. Lawrence River (Murwaski & Pacheco 1977).

Sex ratio and mortality factors

Sex ratios were calculated for fish captured in the Suwannee River and for those captured around the Winyah Bay jetties. Huff's (1975) analysis of the pooled sex data for 1972–1973 did not differ significantly from a 1:1 sex ratio; however, the G statistic was highly significant for seasonal sex ratios. He concluded that the heterogenous sex ratios in the spring and fall are indirect evidence of differential migration routes taken by the pre- and post-spawning fish. In Winyah Bay, male: female ratios of commercially landed fish during spring of 1978 to 1982 ranged from 1:2 to 1:4. Chi-square analysis indicated that the sex ratios were highly significant ($P<0.005$) during each year. Thus, on the average 72% of all fish landed in South Carolina were females (Smith et al. 1984). Reason for the differential sex ratios was not apparent although the use of large mesh gill nets (30.5 to 38.1 cm stretch mesh) may provide some selectivity for females which are typically larger than the males.

Little information exists on the mortality factors which affect Atlantic sturgeon. For fish from the Suwannee River Huff (1975) estimated an annual survivorship of 54% between successive years for sturgeon 8–12 year. In South Carolina, Leland (1968) observed that long-nose gar will attack schools of young sturgeon and Scott & Crossman (1973) reported that the sea lamprey, *Petromyzon marinus*, will attack and kill large sturgeon. Of all factors however, man exerts the greatest influence either directly through fishing pressure or killing of incidentally caught juveniles in shad nets (Leland 1968) as well as indirectly through alteration of critical habitats.

Management

Propagation efforts

The rapid decline in stocks of sturgeon during the last quarter of the nineteenth century was obvious to early fishery managers. Their approach to maintain specific stocks was through artificial propagation (Ryder 1890, Cobb 1900). For example, W. de C. Ravenel, manager for propagation and distribution of food fishes for the U.S. Fish Commission, stated: '... there is no subject in fish culture, excepting the lobster, that we have given more time and thought to in the last few years. (...) We are prepared to do more for the sturgeons than anything else except the lobster' (Stone 1900).

The first successful artificial spawning of Atlantic sturgeon was accomplished on the Hudson River in 1875 by Seth Green and A. Marks with the New York State Fish Commission. The ovaries and testes were removed from ripe fish obtained from fishermen and the eggs and milt artificially mixed. About 40000 young were hatched followed by 60000 the next week. In 1888, the U.S. Fish Commission initiated its artificial propagation studies on the Delaware River under the direction of J. A. Ryder (Ryder 1890). His efforts as well as later efforts were thwarted by two major problems: the acquisition of simultaneously ripe males and females in sufficient numbers, and, fungal infestations (*Achlya* and *Saprolegnia*) of incubating eggs. In spite of the interest and effort to develop artificial propagation techniques, most activities in the United States and Canada were terminated by 1912 (see reviews by Dean 1894, Leach 1920, Harkness & Dymond 1961, Hoff 1980, Smith & Dingley 1984).

Recently, in 1979, efforts to develop spawning techniques for Atlantic sturgeon in South Carolina were successful (Fig. 2). Gametes were obtained using an injection of sturgeon pituitary coupled with surgical removal of ovulated eggs. The artificially inseminated eggs were incubated in McDonald jars (Smith et al. 1980). Additionally, information on the use of ponds for rearing juveniles and to hold brood stock was provided (Smith et al. 1981). This hatchery development effort is

Fig. 2. Photograph of a three month old cultured Atlantic sturgeon (10.1 cm TL).

continuing and in 1983, it was expanded to include shortnose sturgeon. The feasibility of stocking cultured juveniles to support certain sturgeon fisheries in the U.S.S.R. has been demonstrated (Sergei Doroshov, personal communication) but additional biological information should be obtained before attempting such large-scale restocking programs in the United States.

Laws and regulations

Failure of early culture efforts to maintain sturgeon stocks resulted in the establishment of a variety of management regulations. Such regulations included gear restrictions, minimum harvest size, closed seasons, and closed areas. Current regulations range from full protection (e.g. Pennsylvania) to no specific regulations (e.g. Maine) (Table 4). For the most part, the various regulations which do permit harvesting neither are based on sufficient biological and fisheries data to justify such regulations nor are they aimed at long-range rational management of the species. During 1983, all state agencies responsible for management of coastal fishery resources along the Atlantic and Gulf coasts, were surveyed concerning current or planned regulations and/or activities on Atlantic sturgeon (Smith, unpublished). Responses were basically similar in that there were no current or near-term plans for management of this species.

Needs and recommendations

There is definite need for biological studies on the Atlantic sturgeon. Such studies should attempt to: identify and characterize specific spawning and nursery areas; provide information on growth rates and mortality factors; establish parent/progeny relationships; provide information on movements of juveniles and adults to delineate contributions and boundaries of various stocks; and, assess the feasibility of stock rehabilitation efforts based on habitat modifications, management regulations, and through the use of artificial propagation techniques.

Recently, the status of Atlantic sturgeon on a state-by-state basis was provided by Hoff (1980). In only Canada, New York, and South Carolina did he provide a status of 'managed fishery' (Table 5). For the entire range, his summary categorization of Atlantic sturgeon was as a species of 'special concern'. This is defined by Deacon et al. (1979) as '... that species that could become threatened or endangered by relatively minor disturbances to their habitat, or that requires additional information to determine its status and therefore commercial fishing should be discouraged, or, as in this instance severely restricted.' Since Hoff's report, however, the sturgeon landings in South Carolina during 1983 and 1984 have dramatically declined to the point where a total fishing ban is now being implemented. Thus, only Canada and New York would retain his designated status of 'managed fishery'.

There have been management recommendations proposed for Atlantic sturgeon. In South Carolina, Leland (1968) recommended moving the shad fishery further upstream to protect the juvenile sturgeon which primarily inhabit the lower estuary from incidental capture and killing in shad

Table 4. Summary of state regulations on the harvesting of Atlantic sturgeon during 1983.

State	Sturgeon fishery	Laws and regulations
Canada	directed	minimum mesh (stretch) 33 cm; minimum length 122 cm TL; open season except June
Maine	incidental	none
New Hampshire	incidental	none
Massachusetts	incidental	none
Rhode Island	incidental	none
Connecticut	none	prohibited harvest of sturgeon
New York	directed	minimum length 122 cm
New Jersey	incidental	capture by any legal gear type; minimum length 107 cm if sold
Pennsylvania	none	prohibited harvest of sturgeon
Maryland	incidental	minimum weight 11,3 kg
Daleware	incidental	minimum mesh (stretch) 28 cm; minimum length 137 cm
Virginia	incidental	minimum length 51 cm
North Carolina	incidental	none; no gill nets having mesh (stretch) greater than 15 cm can be fishes in ocean from Carolina Beach to S. C. boundary during February 1 to June 30.*
South Carolina	directed	minimum mesh (stretch) 25 cm; nets must be licensed; open season March 1–October 1 except in ocean around Winyah Bay jetties (open February 15 to April 15)
Georgia	directed	minimum gill net mesh (square) 15 cm; open season January 15 to June 30
Florida	directed	minimum mesh (stretch) 25 cm; gill nets only
Alabama	none	none
Mississippi	none	listed as an endangered species
Louisiana	none	none
Texas	none	none

* Enacted to keep S. C. sturgeon fishermen from fishing in N. C.

nets. He also suggested the establishment of 'fish life areas' comparable to those established for land dwelling animals for protecting sturgeon stocks. In New York, the Hudson River Atlantic sturgeon stock has been a focus of study during recent years (Dovel 1978, 1979). Using Hudson River data in conjunction with biological information from other areas Young et al. (1984) modified Dovel's suggestions (Dovel 1979) and developed a series of management recommendations and alternatives for the Hudson River stock based on an age-structured population model. Their recommendations were:

1. to establish a minimum preferred harvest size of 183 cm to permit fish the opportunity to mature and spawn before harvesting;
2. to allow a conservative maximum allowable catch (10 metric tons), until more information on stock size and age structure was available;
3. to require special licensing and reporting requirement for fishermen;
4. to initiate a fishery monitoring program to collect data;
5. to provide a team approach to develop and implement a management plan which would allow periodic evaluation of success.

Such efforts are laudable and the recommendations as indicated by Dovel (1979) could be broadly applied to other sturgeon populations with modifications made based on specific biological and fishery data from local stocks.

Unfortunately, the biological characteristics of this species makes it highly susceptible to long term damage from short term perturbations. Lack of conservative protective measures now may result in the further devastation and/or disappearance of local stocks. Indeed, the Atlantic sturgeon is in need of immediate protection throughout much, if not all, of its range and perhaps the best approach would be the immediate establishment of a total moratorium on exploitation of this species. At present, I would strongly recommend the Federal designation 'threatened' or 'endangered' until such time that the biological data would support the

Table 5. Recommended status of Atlantic sturgeon, especially riverine residents, by state (from Hoff 1980).

State	Recommended status
Canada	managed fishery
Maine	special concern
New Hampshire	threatened
Massachusetts	threatened
Rhode Island	threatened
Connecticut	threatened
New York	managed fishery
New Jersey	special concern
Pennsylvania	special concern
Maryland	special concern
Delaware	special concern
Virginia	threatened
North Carolina	special concern
South Carolina	managed fishery
Georgia	special concern
Florida	threatened/special concern
Alabama	threatened
Mississippi	endangered
Louisiana	special concern
Texas	no classification

rational, managed harvesting of Atlantic sturgeon.

These fish were once an important item of commerce. Perhaps through man's positive intervention the current populations of Atlantic sturgeon could be at least maintained, if not expanded for possible use by future generations.

Acknowledgements

Contribution No. 189 from the South Carolina Marine Resources Center. Preparation of the manuscript was jointly sponsored by the U.S. Fish and Wildlife Service and the State of South Carolina. Karen Swanson prepared the figures.

References cited

Bachus, R.H. 1951. New and rare records of fishes from Labrador. Copeia 1951: 288–294.

Bigelow, H.B. & W.C. Schroeder. 1953. Fishes of the Gulf of Maine. Fish. Bull. 53: 577.

Borodin, N. 1925. Biological observations on the Atlantic sturgeon *Acipenser sturio*. Trans. Amer. Fish. Soc. 55: 184–190.

Cobb, J.N. 1900. The sturgeon fishery of Delaware River and Bay. Rep. U.S. Comm. Fish and Fisheries for 1899, Part 25: 369–380.

Deacon, J.E., G. Kobetich, J.D. Williams & S. Conteras. 1979. Fishes of North America endangered, threatened or of special concern. Fisheries 4: 29–44.

Dean, B. 1893. Notes on the spawning conditions of the sturgeons. Zool. Anz. 16: 473–475.

Dean, B. 1894. Recent experiments in sturgeon hatching on the Delaware River. U.S. Fish Comm. Bull. (1893) 13: 335–339.

Dees, L.T. 1961. Sturgeons. U.S. Fish Wildl. Serv., Fish Leafl. 526. 8 pp.

Dovel, W.L. 1978. Biology and management of shortnose and Atlantic sturgeon of the Hudson River. N.Y.S. Dept. Envir. Cons., Perf. Rep. Proj. AFS-9-R. 181 pp.

Dovel, W.L. 1979. The biology and management of shortnose and Atlantic sturgeon of the Hudson River. N.Y.S. Dept. Envir. Cons., Final Rep. Proj. AFS-9-R. 54 pp.

Galligan, J.P. 1960. History of the Connecticut River sturgeon fishery. Conn. Wildl. Cons. Bull. 6:1, 5–6.

Gent, J.J. 1675. An account of two voyages to New England. A description of the country, native and creatures. *In:* Coll. Mass. Hist. Soc. 3rd ser. III. 1833.

Greeley, J.R. 1937. Fishes of the area with annotated list. pp. 45–103. *In:* A biological survey of the lower Hudson watershed. Suppl. 26th Ann. Rpt. N.Y. Cons. Dept. Biol. Serv. 11. Albany.

Harkness, W.J.K. & J.R. Dymond. 1961. The lake sturgeon, the history of its fishery and problems of conservation. Ontario Dept. Lands, Forests, Toronto. 121 pp.

Hildebrand, S.F. & W.C. Schroeder. 1928. Fishes of Chesapeake Bay. U.S. Bur. Fish. Bull. 43, Pt. 1, Washington. 366 pp.

Hoff, J.G. 1980. Review of the present status of the stocks of the Atlantic sturgeon *Acipenser oxyrhynchus* (Mitchill). Southeast. Mass. Univ., Rpt. to Nat. Mar. Fish. Serv, North Dartmouth. 136 pp.

Holland, B.F. Jr. & G.F. Yelverton. 1973. Distribution and biological studies of anadromous fishes offshore North Carolina. N.C. Dept. Nat. Econ. Res. Spec. Sci. Rpt. 24, Morehead City. 132 pp.

Hoover, E.E. 1938. Biological survey of the Merrimack watershed. Fish Game Comm., Concord. 238 pp.

Huff, J.A. 1975. Life history of Gulf of Mexico sturgeon, *Acipenser oxyrhynchus desoti*, in Suwannee River, Florida. Fla. Dept. Nat. Res., Mar. Res. Publ. 16, St. Petersburg. 32 pp.

Jones, P.W., F.D. Martin & J.O. Hardy. 1978. Development of fishes of the mid Atlantic Bight. Acipenseridae through Ictaluridae. Fish Wildl. Serv. Biol. Prog. FWS/OBS – 78/12, Washington, D.C. 366 pp.

Knox, J.D. & M.J. Dadswell. 1980. An annotated bibliography on reproduction and early life history of sturgeon (Osteichthyes: Acipenseridae) with reference to fish passage and artificial culture in the U.S.S.R. Northeast Util. Serv. Co., Hartford. 114 pp.

Leach, G.C. 1920. Artificial propagation of sturgeon, review of sturgeon culture in the United States. Rep. U.S. Fish Comm. 1919: 3–5

Leland, J.G., III. 1968. A survey of the sturgeon fishery of South Carolina. Contr. Bears Bluff Lab. 47: 1–27.

Magnin, E. 1962. Rechearches sur la systematique et la biologie des Acipensendes, *Acipenser sturio* L., *Acipenser oxyrhynchus* Mitchill et *Acipenser fulvescens* Raf. Ann. Sta. Cent. Hydrobiol. Appl. 9: 7–242.

Magnin, E. 1964. Croissance en longeur de trois esturgeons d'Amérique du Nord: *Acipenser oxyrhynchus* Mitchill, *Acipenser fulvescens* Raffinesque, et *Acipenser brevirostris* LeSueur. Verh. int. Verein. theor. angew. Limnol. 15: 968–974.

Mansueti, A.J. & J.D. Hardy. 1967. Development of fishes of the Chesapeake Bay Region. Nat. Res. Inst. Univ. of Maryland. Solomons. 202 pp.

Miller, R.R. 1972. Threatened freshwater fishes of the United States. Trans. Amer. Fish. Soc. 101. 239–252.

Murawski, S.A. & A.L. Pacheco. 1977. Biological and fisheries data on Atlantic sturgeon, *Acipenser oxyrhynchus* (Mitchill). Nat. Mar. Fish. Ser., Tech. Ser. Rep. 10: 1–69.

Rulifson, R.A. & M.T. Huish. 1982. Anadromous fish in the southeastern United States and recommendations for development of a management plan. U.S. Fish Wild. Ser., Fish. Resour. Reg. 4. 525 pp.

Ryder, J.A. 1890. The sturgeon and sturgeon industries of the eastern coast of the United States, with an account of experiments bearing upon sturgeon culture. U.S. Fish Comm., Bull. (1888) 8: 231–238.

Scott, W.B. & E.J. Crossman. 1973. Freshwater fishes of Canada. Fish. Res. Board Can. Bull. 184. 966 pp.

Smith, H.M. 1907. The fishes of North Carolina. N.C. Geol. Econ. Surv., Vol. II. 453 pp.

Smith, T.I.J. & E.K. Dingley. 1984. Review of biology and culture of Atlantic (*Acipenser oxyrhynchus*) and shortnose sturgeon (*A. brevirostrum*). J. World Maricul. Soc. 15 (in press).

Smith, T.I.J., E.K. Dingley & D.E. Marchette. 1980. Induced spawning and culture of Atlantic sturgeon. Prog. Fish Cult. 42: 147–151.

Smith, T.I.J., E.K. Dingley & D.E. Marchette. 1981. Culture trials with Atlantic sturgeon, *Acipenser oxyrhynchus* in the U.S.A. J. World Maricul. Soc. 12: 78–87.

Smith, T.I.J., D.E. Marchette & R.A. Smiley. 1982. Life history, ecology, culture and management of Atlantic sturgeon, *Acipenser oxyrhynchus oxyrhynchus* Mitchill, in South Carolina. S.C. Wildl. Mar. Resour. Res. Dept., Final Rep. to U.S. Fish. Wildl. Ser. Proj. AFS-9. 75 pp.

Smith, T.I.J., D.E. Marchette & G.F. Ulrich. 1984. The Atlantic sturgeon fishery in South Carolina. N. Amer. J. Fish. Mgmt. 4: 164–176.

Squires, T.S. & M. Smith. 1979. Distribution and abundance of shortnose and Atlantic sturgeon in the Kennebec River estuary. Maine Dept. Mar. Resour., Compl. Rep. to U.S. Fish Wildl. Ser., Proj. AFC-19. 51 pp.

Stone, L. 1900. The spawning habits of the lake sturgeon (*Acipenser rubincundus*). Trans. Amer. Fish Soc. 29: 118–128.

Van Den Avyle, M.J. 1983. Species profiles: life histories and environmental requirements (South Atlantic) – Atlantic sturgeon. U.S. Fish Wildl. Ser., Div. Biol. Ser. FWS/OBS-82/11. U.S. Army Corps Eng., TREL-82-4. 38 pp.

Vladykov, V.D. & J.R. Greeley. 1963. Order Acipenseroidei. Fishes of the Western North Atlantic. Sears Found. Mar. Res., Yale University, New Haven. 630 pp.

Wood, W. 1634. New England's prospect. London. (from Murwaski and Pacheco 1977).

Wooley, C.M. 1985. Evaluation of morphometric characters used in taxonomic separation of Gulf of Mexico sturgeon, *Acipenser oxyrhynchus desotoi*. pp. 97–103. *In:* F.P. Binkowski & S.I. Doroshov (ed.) North American sturgeons: biology and aquaculture potential, Dev. EBF 6, Dr W. Junk Publishers, Dordrecht (this volume).

Wooley, C.M. & E.J. Crateau. 1982. Observations of Gulf of Mexico sturgeon (*Acipenser oxyrhynchus desotoi*) in the Applachicola River, Florida. Fla. Sci. 45: 244–248.

Young, J.R., T.B. Hoff, W.P. Dey & J.G. Hoff. 1984. Management recommendations, for a Hudson River Atlantic sturgeon fishery based on an age-structured population model. Symp. Hudson R. Fishes, 1981, Hudson R. Envir. Soc. N.Y. 27 pp. (in press).

Received 15.5.1984   Accepted 16.1.1985

The lake sturgeon, *Acipenser fulvescens*, in the Menominee River, Wisconsin-Michigan

Thomas F. Thuemler
Wisconsin Department of Natural Resources, Box 16, Industrial Parkway, Marinette, WI 54143, U.S.A.

Keywords: Population estimates, Age, Growth, Exploitation, Creel census, Harvest

Synopsis

The Menominee River is the boundary between the Upper Peninsula of Michigan and northeastern Wisconsin. It contains one of the few fishable lake sturgeon populations remaining in either state. Two sections of the river harbor naturally reproducing sturgeon stocks. Dams at either end of these sections curtail upstream movement of fish; however, some sturgeon move downstream over the dams. Surveys conducted in 1969 and 1970 and again in 1978 and 1979 intensively studied the sturgeons in a 42 km stretch of the river. Mark and recapture estimates of the number of sturgeon longer than 25 cm in this section ranged from 1884 to 2865. The number of fish longer than 107 cm ranged from 185 to 243. An estimate of 2834 sturgeon between 25 and 165 cm was made on a different 34 km stretch in 1979. Censuses on the hook and line fishery in 1969 and 1970 estimated sturgeon harvests of 59 and 48 fish, respectively. In 1974, the size limit was raised from 107 to 127 cm. Harvests of these larger fish were estimated at 19 and 41 in 1981 and 1982, respectively. Some other characteristics of the sturgeons and the fishery on the Menominee are also included.

Introduction

The Menominee River forms the border between northeastern Wisconsin and the Upper Penisula of Michigan (Fig. 1). It contains one of the few fishable lake sturgeon, *Acipenser fulvescens*, stocks remaining in either state. The objective of this study was to describe the characteristics of the lake sturgeon and its fishery in order to provide proper management. Management is aimed at preservation of this stock while still providing a recreational fishery and reestablishing the sturgeon in an area of the Menominee River where it has been eliminated in the past. The management of the Menominee River sturgeon is a joint effort between the states of Wisconsin and Michigan. Much of the early work on this river was conducted by Priegel (1973).

Methods

The Menominee River is formed by the Brule and Michigammie rivers and then flows for roughly 152 km before entering the waters of Green Bay at the cities of Marinette, WI and Menominee, MI (Fig. 1). It is a hardwater river having alkaline, lightly stained water. There are ten dams throughout its length which regulate flow. The mean annual discharge in the study area is 113280 liters per second (Holmstrom et al. 1983) and has an average width of 144 m. In addition to lake sturgeon, the river contains walleye, *Stizostedion vitreum*, northern pike, *Esox lucius*, largemouth bass, *Micropterus salmoides*, smallmouth bass, *Micropterus dolomieui*, channel catfish, *Ictalurus punctatus*, rock bass, *Ambloplites rupestris*, yellow perch, *Perca*

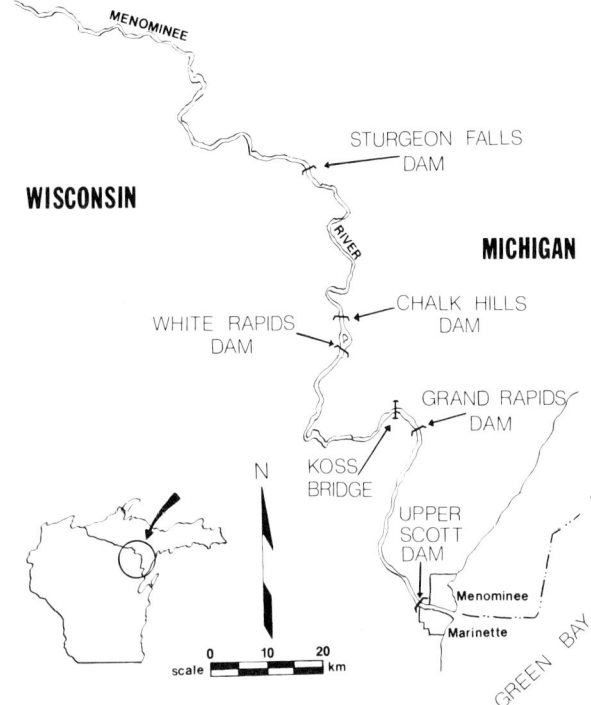

Fig. 1. Lake sturgeon study area on the Menominee River, Wisconsin – Michigan boundary water.

flavescens, white sucker, *Catostomus commersoni,* carp, *Cyprinus carpio,* redhorse, *Moxostoma* spp., and a large variety of course fish.

Sturgeon were sampled in the river using four A.C. boomshocker electrofisher units moving abreast downstream (Novotny & Priegel 1974). Large specially designed nets were used to capture the sturgeon (Folz et al. 1983). Dam operators kept water levels as low as possible during the sampling, but some of the holes in the river remained 4 to 5 m deep and were extremely hard to electrofish effectively. All the sturgeon captured were measured to the nearest 1.3 cm total length. A pectoral fin spine was removed for aging purposes. Mark and recapture density estimates were made on lake sturgeon in various stretches of the river using the Peterson formula as modified by Chapman (1951). All sturgeon 25 cm and larger were tagged with either a Monel metal tag in the dorsal fin or a Floy dart tag in the back or caudal fin. Some of the sturgeon were marked both ways and retention was found to be much better with the Monel tags.

Age was determined for 1464 sturgeon from the Menominee River by counting the annuli on a thin cross section of the pectoral fin spine (Priegel 1973).

A census of anglers was conducted on the 42 km stretch of river between the White Rapids dam and the Koss bridge in 1981 and 1982 following the methods described by Priegel (1973). The sturgeon fishing season started on September 5, and ran through November 1, in 1981 and on September 4 through November 1 in 1982.

Results

Vital statistics of the sturgeon population

Historically, sturgeon were present in the Menominee upstream to Sturgeon Falls (Fig. 1). However, in recent studies we have found no sturgeon above the White Rapids dam. The stretch of river between the White Rapids and Grand Rapids dams contains sturgeon, and sturgeon are also found between the Grand Rapids dam and the Upper Scott dam in the City of Marinette. Our tagging studies show that there is very little mixing between these two groups. A few sturgeon have moved downstream over the dams, however, there is no way that sturgeon can move upstream past these structures.

Most of our survey work has been done on the 42-km middle section of the river between the White Rapids dam and Grand Rapids dam. A density estimate on all lake sturgeon longer than 25 cm was made on this stretch of river in 1978 and compared to estimates made by Priegel (1973) (Table 1). Sturgeon, 12 to 15 cm, were also found in the river, but these were not included in the estimates. Although there were wide confidence intervals on these estimates, all three fell in the same range of roughly 2000 fish shorter than 107 cm and 200 longer than 107 cm. Quite possibly this is an underestimate of the larger fish, as they are found in the deepest holes in the river. Length frequencies of the sturgeon stock from 1969 to 1978 are given in Figure 2. There is very little difference in the size structure between these two surveys, especially for the larger fish.

A 34 km stretch of river between the Grand Rapids dam and the Upper Scott dam in Marinette was also surveyed by Priegel (1973) in 1970 and repeated in 1979. The 1979 estimate of all sturgeon between 25 and 165 cm in this section was 2834 fish with a confidence interval of ± 39%. Length frequencies of sturgeon taken in single run surveys in 1970, 1979 and 1983 are given in Figure 3. Size structure of these three groups is very similar, especially among fish greater than 100 cm.

Growth of 1464 Menominee River fish was somewhat slower than that of Lake Winnebago sturgeon (Priegel & Wirth 1975), especially in their early years of life (Fig. 4). The Menominee River fish are faster growing, however, than lake sturgeon in Canadian waters (Cuerrier & Roussow 1951, Harkness 1923, Roussow 1957). The average Menominee River sturgeon does not reach the legal size of 127 cm until it is 20 years old, and at that time it weighs 15 kg.

The lake sturgeon fishery

The only legal method to harvest lake sturgeon from the Menominee River is by hook and line during a two month fall season. There was no open season prior to 1946 (Table 2). Since that time the season has gradually become more restrictive. The 1983 season was the first in which all sturgeon taken had to be tagged and registered at a Department of Natural Resources station.

A creel census was conducted on the section of river between the White Rapids dam and the Koss bridge during the 1981 and 1982 seasons. The information obtained was compared to similar censuses run by Priegel (1973) during the 1969 and 1970 seasons (Table 3). A 107 cm size limit was in effect during the first two years of the census. In 1974 the size limit was raised to 127 cm and this limit was in

Table 1. Density estimates of lake sturgeon in a 42 km stretch of the Menominee River, White Rapids dam to Koss bridge.

Size (cm)	Year	Estimate	95% confidence intervals
25–107	1969	1641	± 35%
	1970	2680	± 19%
	1978	2543	± 26%
Larger than 107	1969	243	± 71%
	1970	185	± 30%
	1978	206	± 44%

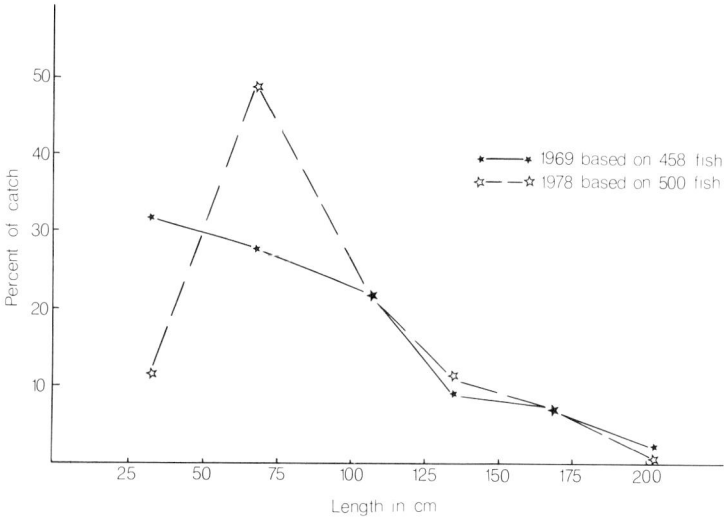

Fig. 2. Length frequencies of lake sturgeon, in the White Rapids to Grand Rapids section of the Menominee River, Wisconsin 1969 and 1978.

effect during the 1981 and 1982 seasons. Fishing pressure dropped by roughly one-half in the later two years of the census. I can only surmise that the increased size limit and the development of a tremendous salmonid fishery on Green Bay contributed to the decrease in pressure. The number of legal sturgeon taken dropped after the increased size limit went into effect, but it did not drop as much as expected. During the 1982 season, there was one-half the fishing pressure that there was in 1970, the larger size limit was in effect and still only seven fewer sturgeon were taken in 1982 than in 1970. Even in years with good fishing conditions, like 1982, it is only the extremely patient or lucky angler who catches a legal fish. There was only one legal fish harvested that year for every 154 hours of

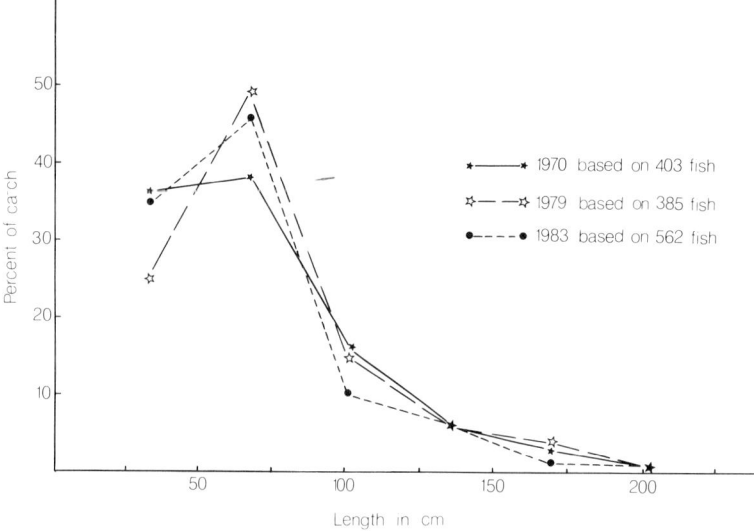

Fig. 3. Length frequencies of lake sturgeon in the Grand Rapids dam to Upper Scott Flowage section of the Menominee River, Wisconsin 1970, 1979 and 1983.

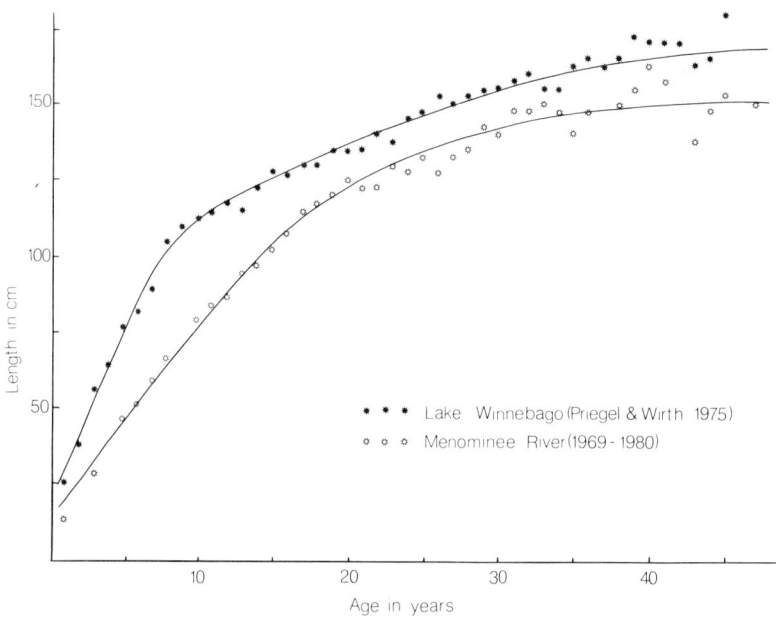

Fig. 4. Age-length relationships of lake sturgeon from the Menominee River and Lake Winnebago, Wisconsin.

Table 2. History of the sturgeon fishing regulations on the Menominee River.

Year	Season length	Bag limit	Size limit (cm)
Prior to 1946	No open season		
1946	1st Sat. in Sept to Oct 15	5 per season	76
1953	1st Sat. in Sept to Oct 15	3 per season	76
1959	2nd Sat. in Sept to Oct 15	3 per season	76
1963	1st Sat. in Sept to Oct 15	2 per season	107
1966	1st Sat. in Sept to Nov 1	2 per season	107
1974	1st Sat. in Sept to Nov 1	2 per season	127
1983	Mandatory registration and tagging required		

fishing. In other years that figure was as low as one legal fish in 332 hours. Anglers do catch quite a few sublegal sturgeon and the catch rate on these sublegals is increasing from the early years of the census. Veteran anglers verify that they are catching more sublegal fish now than they have in the past.

In the White Rapids to Grand Rapids section of the river, exploitation was determined by Priegel (1973) to be 13% in 1969 and 17% in 1970. These percentages were based on limited returns of marked fish to the creel. If these rates are accurate, they are much too high and this is the reason the size limit was increased to 127 cm in 1974. Our policy in Wisconsin is to manage for a 5% or less exploitation rate on sturgeon. Exploitation data is not available for the 1981 and 1982 seasons, but Figures 2 and 3 show very little difference in the length frequency of sturgeon captured in our surveys over the years. If overharvest were occurring, I would expect to see a decline in the percentage of large fish in the population.

The average size of sturgeon in the harvest in 1981 and 1982 was 139 cm. The Wisconsin Department of Natural Resources has a policy that at least 50% of the sturgeon in the harvest be 13 cm over the minimum size of 127 cm to assure against overharvest. This condition was met in the fishery on the Menominee River during the 1981 and 1982 seasons.

Two of the problems we have in getting reliable estimates of exploitation on sturgeon in the Menominee River are: (1) The difficulty in making statistically sound density estimates, especially for the larger sturgeon; and (2) the small number of legal size sturgeon actually seen in the harvest each year. The mandatory registration system that went into effect in 1983 will provide a much better account of the harvest than our creel estimates have in the past.

Repopulation efforts

Sturgeon Falls was historically the farthest point upstream that sturgeon were found in the Menominee River. However, surveys since 1969 have revealed no sturgeon above the White Rapids dam (Fig. 1). Habitat in this section of the river appears well suited for sturgeon and the river in general is quite similar to the lower reaches where sturgeon are now present. Pollution could have been respon-

Table 3. Estimated annual harvest of lake sturgeon from the Menominee River, White Rapids dam to Koss bridge section.

	1969	1970	1981	1982
Legal lake sturgeon taken*	59	48	19	41
Harvest rate per hour	0.004	0.004	0.003	0.007
Sublegal lake sturgeon caught	1075	1420	1236	1553
Sublegals per hour	0.075	0.129	0.193	0.248

* Size limit was changed from 107 to 127 cm in 1974.

sible for the demise of sturgeon in the upper Menominee, but water quality has been greatly improved in this area in recent years. Other gamefish such as walleye, smallmouth bass and northern pike have increased their numbers.

An effort is now being made to try and reestablish the sturgeon population. Lake sturgeon eggs taken from fish on the Wolf River in central Wisconsin (Folz et al. 1983) were incubated, hatched and raised at the Wild Rose State Fish Hatchery in Wild Rose, Wisconsin (AveLallemant et al. 1983). In October 1982, 290 yearling lake sturgeon averaging 18 cm in length and 18 g in weight were stocked in the Sturgeon Falls section of the Menominee River. All of these fish were marked with coded wire tags. In June of 1983, 11000 lake sturgeon juveniles averaging 30 mm in length also were stocked into this section. Future electrofishing surveys will be conducted on this 32 km section of the Menominee to determine the survival of these hatchery reared sturgeon.

Acknowledgments

I wish to thank Gordon Priegel and Milton Burdick of the Wisconsin Department of Natural Resources and Lud Frankenberger and Buddy Jacob of the Michigan Department of Natural Resources who coordinated the initial surveys and assisted in much of the survey work; Greg Kornely who coordinated the creel census in 1981 and 1982 and analyzed the data; and Tina Oman who drafted the figures for the manuscript.

References cited

AveLallemant, S., D. Czeskleba & T. Thuemler. 1983. Artificial spawning and rearing of lake sturgeon at the Wild Rose State Fish Hatchery, Wisconsin. Fish Culture Note 5, Wisconsin Department of Natural Resources, Madison. 7 pp.

Chapman, D.G. 1951. Some properties of the hypergeometric distribution with applications to zoological censuses. Univ. Calif. Publ. in Statistics 1: 131–160.

Cuerrier, J.P. & G. Roussow. 1951. Age and growth of lake sturgeon from Lake St. Francis, St. Lawrence River. Report material collected in 1947. Canad. Fish. Cult. 10: 17–29.

Folz, D.J., D.G. Czeskleba & T.F. Thuemler. 1983. Artificial Spawning of Lake Sturgeon in Wisconsin. Prog. Fish-Cult. 45: 231–232.

Harkness, W.J.K. 1923. Rate of growth and food of lake sturgeon (*Acipenser rubicundus*). Publ. Ont. Fish Res. Lab. 18: 15–42.

Holmstrom, B.K., C.A. Han & R.M. Erickson 1983. Water Resources Data, Wisconsin Water Year 1982. U.S. Geological Survey Water Data Report WI-82-1. 426 pp.

Novotny, D.W. & G.R. Priegel. 1974. Electrofishing boats. Technical Bulletin 73, Wisconsin Department of Natural Resources, Madison. 48 pp.

Priegel, G.R. 1973. Lake sturgeon management on the Menominee River. Technical Bulletin 67, Wisconsin Department of Natural Resources, Madison. 20 pp.

Priegel, G.R. & T.L. Wirth. 1975. Lake sturgeon harvest, growth and recruitment in Lake Winnebago, Wisconsin. Technical Bulletin 83, Wisconsin Department of Natural Resources, Madison. 24 pp.

Roussow, G. 1957. Some considerations concerning sturgeon spawning periodicity. J. Fish. Res. Board Can. 14: 553–572.

Received 15.5.1984   Accepted 14.2.1985

Artificial spawning and rearing of lake sturgeon, *Acipenser fulvescens*, in Wild Rose State Fish Hatchery, Wisconsin, 1982–1983

Donald G. Czeskleba[1], Steven AveLallemant[1] & Thomas F. Thuemler[2]
[1] *Wisconsin Department of Natural Resources, Wild Rose State Fish Hatchery, Wild Rose, WI 54984, U.S.A.*
[2] *Wisconsin Department of Natural Resources, Box 16, Industrial Parkway, Marinette, WI 54143, U.S.A.*

Keywords: Propagation, Juveniles, Larvae, Embryos, Eggs, Survival, Diet, Insemination, Hatching, Development, Stocking

Synopsis

Attempts to culture lake sturgeon, *Acipenser fulvescens*, in the past have generally met with limited success. The Wisconsin Department of Natural Resources has been experimenting with artificial propagation of this species since 1979. The intent has been to develop egg collection and handling techniques, hatching regimes, larva and juvenile diet formulations, and to evaluate juvenile survival after stocking. Eggs were collected by caesarian section and fertilized with milt from ripe males taken during annual sturgeon spawning runs on the Fox and Wolf rivers in central Wisconsin. After insemination, the eggs were treated in a saturated solution of Bentonite clay and transported to the hatchery. Eggs were incubated at temperatures ranging from 13–16° C and embryos began hatching within 4 to 8 days. Hatching success ranged from 42 to 96%. Yolksac absorption was complete within 10 days of hatching. Larvae then became positively phototactic and swam actively as if searching for food. Successful larval diets consisted of live brine shrimp nauplii followed by larger zooplankton, primarily *Daphnia* sp. Juveniles grew best on diets of live *Tubifex* sp. and chopped earthworms. Liver, fish mash (ground up trout) and pelleted dry food were poorly accepted. Hatchery reared sturgeon grew more slowly than did wild fish.

Introduction

Declines in sturgeon densities across the United States have stimulated interest in development of methods for artificial propagation of these fish. Attempts to culture sturgeon in the United States have historically met with very limited success (Harkness & Dymond 1961).

The Wisconsin Department of Natural Resources (WDNR) has attempted to artificially spawn and rear lake sturgeon, *Acipenser fulvescens*, since 1979 at the Wild Rose State Fish Hatchery. Cooperative research into lake sturgeon propagation has also been ongoing at the Center for Great Lakes Studies at the University of Wisconsin-Milwaukee. This paper documents the methods developed for artificial propagation of lake sturgeon, including egg collection and handling techniques, hatching regimes, early development, larva and juvenile diet formulation, and tagging techniques to evaluate juvenile survival after stocking.

Methods

Egg collection

Lake sturgeon eggs and milt were collected on April 26, 1982 and on April 27, 1983 on the Fox

F.P. Binkowski & S.I. Doroshov (ed.), North American sturgeons. ISBN 90-6193-539-3
© 1985, Dr W. Junk Publishers, Dordrecht. Printed in the Netherlands. Developments in EBF 6.

River at Eureka, Wisconsin, 37 kilometers upstream from its confluence with Lake Winnebago during annual tagging operations by WDNR personnel (Fig. 1). River temperature was 13° C in 1982 and 14.5° C in 1983. Active spawning was in progress.

Fish were captured with large dip nets as they moved inshore to spawn and were transferred to a stretcher with a pouch for the head where fresh water was poured over the gills. The abdomens of male fish were palpated to express milt into the depression around the vent. The milt was then drawn up with a syringe and placed on ice in glass vials until ripe females could be captured. A total of six females were captured during 1982 and 1983 from which eggs were removed by caesarian section through 35 to 50 mm incision just anterior and to the side of the vent. A speculum was used to spread the incision. After removing the eggs, the incision was closed with dissolving sutures, using triangular shaped cutting needles. Females were kept out of the water from 13–19 min, after which they appeared to recover fully. Males were held for a shorter period, and all were released unharmed.

Eggs were placed in 4.5 l plastic buckets and immediately inseminated with milt collected from at least two males. Milt not contaminated by water was held on ice from 15 min to 2.5 h prior to use. After stirring the eggs and milt, fresh water was added and the bucket was allowed to stand for several minutes. Eggs were immersed and swirled for approximately 5 min in a saturated bentonite clay suspension to reduce egg adhesion and prevent clumping. Excess bentonite was rinsed from the eggs before they were placed in buckets in a foam lined box and taken to the Wild Rose Hatchery.

Egg incubation

Fertilized eggs arrived at the hatchery within 3 h of spawning. They were screened through 4 mm mesh nylon to separate egg clumps and given another coating of clay, after which they were put up in McDonald hatching jars. The eggs from each female were divided in half during 1982; one half was incubated at 13° C and the other half at 16° C. All eggs collected in 1983 were incubated at 13° C.

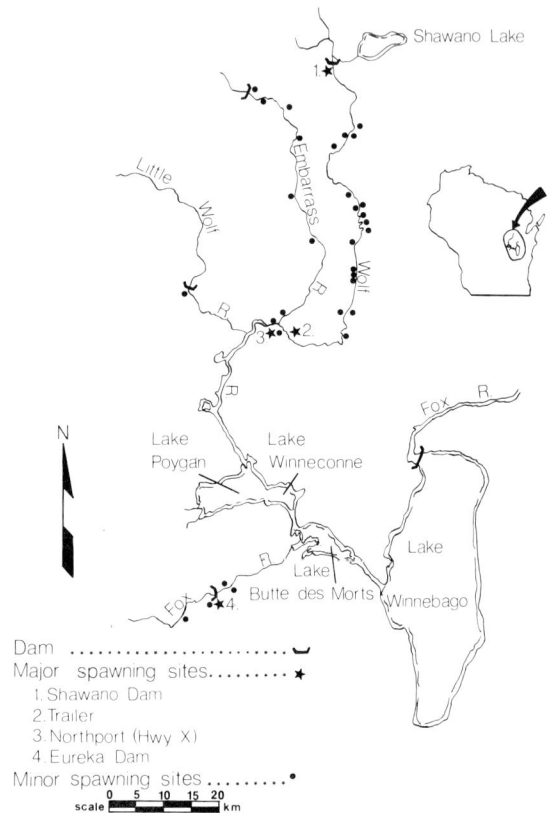

Fig. 1. Lake sturgeon study area on the Lake Winnebago system, Wisconsin.

Water flow was approximately 7.5 l per minute through each jar. Eggs were rescreened after 3 days to remove those coated with fungus and then treated daily with formalin (1:600) until hatching began. Diameter of eggs from each female used in 1982 were measured as uninseminated eggs and at 4 days after fertilization at both incubation temperatures. Newly hatched embryos were allowed to swim out of each hatchery jar into separate 75 l fiberglass washtubs in which rolled astroturf type mats were provided for shelter.

Rearing

After hatching, embryos were reared in 75 l fiberglass wash tubs. Water flow through the tubs was increased to 15 l per minute, while the temperature was held at 20° C in 1982 and 17° C in 1983. Embryos were examined daily for changes in morphological development including total length and allowed to hide in astroturf mats until yolksacs were absorbed. Larvae were measured after yolk-sac absorption. In 1982, continuous feeders were set up on each tub and dry feed (W-14 starter for one tank and W-16 starter for the other) (U.S. Fish & Wildlife Service diets) was offered along with live brine shrimp nauplii, *Artemia* sp. Only live brine shrimp nauplii were offered to one tub of larvae in 1983; in another tub 1000 larvae were offered only Oregon Moist starter pelleted food as a trial.

Larvae developed gas bubbles in the gut by day 21 in 1982. They were switched from the tubs to two large raceways (0.6 × 0.6 × 8.5 m) to provide greater water volume. Water flow was approximately 48 l per minute. Fish in one raceway were fed dry feed only (W-14 #1) (U.S. Fish & Wildlife Service diet) while fish in the other were provided life food consisting of brine shrimp nauplii and *Daphnia* sp. High mortalities and poor condition of the larvae in the raceway receiving dry feed prompted addition of live food as a supplement. In 1983, chopped *Tubifex* sp. were also offered in addition to the *Daphnia* sp. and brine shrimp nauplii.

Small fingerlings were combined and moved to a single raceway at 52 days in 1982 and at 31 days in 1983. Water flow varied between 38 and 46 l per minute. Daily maximum water temperatures varied from 16.7–22.8° C. A variety of food items were offered, including W-14 dry feed, frozen adult brine shrimp, liver puree, a paste made from a combination of these three items, frozen *Daphnia* sp. and chopped earthworms. In 1983 live *Tubifex* sp. was fed when it was available, as was a fish mash (ground up trout) combined with ground *Tubifex* sp. as an attractant. Larvae and juveniles were treated prophylactically at least weekly with chloramine-T-hydrate at concentrations of 10 to 16 milligrams per liter.

Tagging

Surviving sturgeon juveniles were tagged with coded wire tags on August 24, 1982. Also tagged were 300 yearling lake sturgeon received from the Center for Great Lakes Studies, University of Wisconsin-Milwaukee. The tagging method consisted of implanting a small (1 mm) section of wire in the fish's nose. Wires were coded by scoring on four sides and magnetized to permit electronic detection at a later date. Tagged fish were returned to tanks for several days, after which a check was run to determine tag retention. All untagged fish were retagged.

Results

Egg incubation

A total of 326 800 eggs were taken from three female lake sturgeon on April 26, 1982. Fungus problems on eggs despite formalin treatments resulted in the loss of approximately 133 000 eggs. In an overall hatching success of 59.2%, a total of 193 550 embryos emerged from egg envelopes.

Eggs incubated at 13° C had a hatching of 63.5% (range 42–93%) (Table 1). These embryos began hatching after 8 days, an average of 194 accumulated temperature units (TU). One temperature unit is defined as 0.6° C above 0° C for 24 h (Piper et al. 1982). Hatching was complete in just over 12

Table 1. Percent hatching success of lake sturgeon incubated at two different temperatures.

Year	Temperature (°C)	Length of female (cm)					
		163	165	170	171	175	Average
1982	13	42	93		65		63.5
1982	16	42	66		63		55
	Average–1982	42	80		64		59
1983	13			96	62.7	62	65.4

days (286 TU) (Table 2). Embryos had well developed eye spots, mouths and rudimentary development of the dorsal fin at the time of hatching.

Eggs incubated at 16°C had a hatching success rate of 55% (range 42–66%) (Table 1). Hatching began after 4 days (141 TU) and was complete by 6 days (169 TU) (Table 2). Embryos that hatched at 16°C had no eye spots, poorly developed mouths, and no dorsal fin development.

Diameter of uninseminated eggs from the three fish ranged from 3.0–3.5 mm. Fertilized eggs incubated at 13°C ranged from 3.5–4.0 mm in diameter after 4 days while diameter of eggs incubated at 16°C ranged 4.0–4.5 mm (Table 3).

Harkness & Dymond (1961) reported that lake sturgeon incubated at 15.6–17.8°C began to hatch at 141.8 T.U. and hatching was complete by 220.4 TU. Diameter of unfertilized eggs averaged 3.3 mm.

In 1983, 242 700 eggs were taken from 3 lake sturgeon on April 27; 70 000 of these eggs were transferred to other facilities. Hatching totaled 113 000 embryos, an overall hatching success of 65.4% (range 62–96%) (Table 1). Eggs began hatching after 9 days, an average of 230 TU. Hatching was complete between 11 and 12 days (272 TU) (Table 2).

Rearing

Newly hatched embryos showed negative phototaxis and remained buried in astroturf mats provided for cover. Their eyes were poorly pigmented and showed no mouth or fin development. Next day eyes were well pigmented. Infolding for mouth development began the same day and appeared well developed by day 5 after hatching. Dorsal fin rays began to develop on day 3 and began to protrude by day 6. Barbels and pectoral fins were evident by day 7. Pectoral, pelvic and dorsal fins and barbels were well developed by 15 days after hatching. Yolksac absorption took between 8 and 10 days. After 10 days larvae became positively phototactic and swam actively as if searching for food. A black intestinal plug observed just posterior to the yolksac in newly hatched embryos had moved to the vent area in 10 days and was lost in most by day 12. Forty percent of the embryos had begun feeding on brine shrimp nauplii by day 12 and 100% by day 15. Mortality during the free embryo interval was less than 0.5%. Harkness & Dymond (1961) observed yolksac absorption in nine days and a change from negative to positive phototaxis seven days after hatching in lake

Table 2. Temperature units needed to initiate and complete ( ) hatching of lake sturgeon.

Year	Incubated at (°C)	Length of female (cm)					
		163	165	170	171	175	Average
1982	13	207 (283)	179 (280)		196 (294)		194 (286)
1982	16	131 (174)	155 (169)	131 (164)		141 (169)	
1983	13			216 (270)	228 (270)	246 (276)	230 (272)

Table 3. Mean diameter (mm) of lake sturgeon eggs taken from three females and incubated at two different temperatures. (50 eggs measured from each batch, 1982).

Length of female (cm)	Unfertilized eggs	Four-day old eggs at	
		13°C	16°C
163	3.5	4.0	4.5
165	3.0	3.5	4.0
171	3.5	4.0	4.5

sturgeon embryos hatched at 15.6–17.8° C.

In 1982 sturgeon young showed little interest in dry feed at 10 days, although 30% of those examined ate live brine shrimp nauplii by 15 days and 70% by day 23. Brine shrimp nauplii could not be supplied in amounts needed. Mortalities increased up to day 36. Most of the larvae appeared to be thin and weak. An abundant source of larger live food, *Daphnia* sp., seined from hatchery ponds, was offered at 36 days, and was readily accepted. Growth increased and condition improved greatly to day 50 when live *Daphnia* sp. could no longer be obtained. Frozen *Daphnia* sp., frozen adult brine shrimp, liver and dry feed were not well accepted and condition declined. High mortalities of smaller larvae reduced numbers to less than 1000 fish by day 70. The remaining large juveniles were offered chopped earthworms, which they readily accepted. Dry feed, liver and frozen *Daphnia* sp. were offered but were poorly accepted. Growth rate and condition improved again and continued to improve until day 138 when the fish were stocked.

Several juveniles were held in a display aquarium at 22° C and fed live *Tubifex* sp. These fish grew faster than any of those held in tanks at water temperatures that varied from 17–23° C and fed other diets.

Embryos grew quickly during their first 10 days, as the yolksac was absorbed, and average total length (TL) increased from 11.5 to 20.1 mm. Growth rate than declined to day 30 when larvae averaged 24.6 mm TL. An increase in growth to day 50 (36.0 mm TL) was stimulated by a large supply of live *Daphnia* sp., which was exhausted after this date.

The increase in mean total length from day 50 to day 70 (36.0–67.6 mm) was attributed to the death of most of the smaller, weaker fish. The remaining fish grew to 101.4 mm TL by day 123 and experienced low mortality on a diet of chopped earthworms and frozen adult brine shrimp (Fig. 2, 3). Wild juveniles captured in the Wolf River during 1982 grew much faster than hatchery reared juveniles (J. J. Kempinger, unpublished data) (Fig. 2). Wolf River fish were collected in bimonthly electrofishing surveys. Age was estimated based on knowing when the hatching in the river peaked.

Sturgeon in 1983 on Oregon Moist pellets did not appear to accept it, and all died by day 29.

Growth of young to day 40 was better during

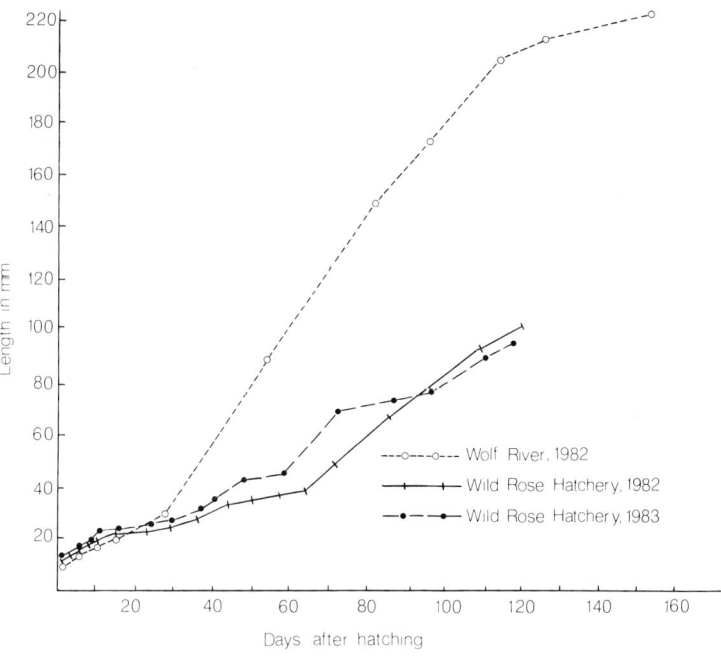

Fig. 2. Comparison of growth of lake sturgeon at the Wild Rose Hatchery with wild fish from the Wolf River.

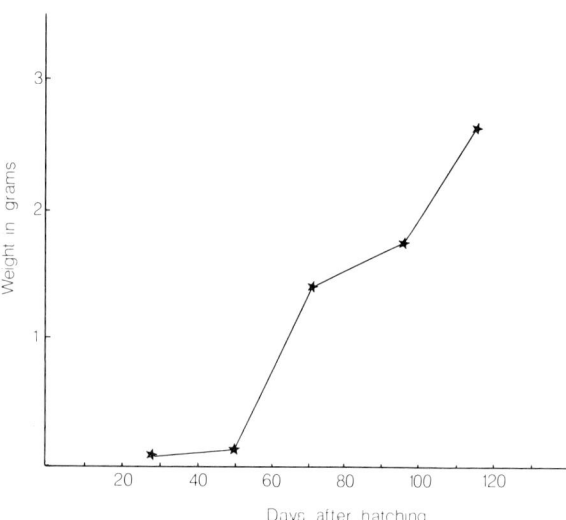

Fig. 3. Age-weight relationship of lake sturgeon raised at the Wild Rose fish hatchery, 1982–1983.

1983 than 1982 (Fig. 2, 3). Faster early growth was likely due to the earlier offering of *Daphnia* sp. and *Tubifex* sp. in 1983. Faster growth during 1983 from day 40 on was attributable to sorting and stocking of smaller juveniles on day 43.

Average total length at stocking (115–120 days) was similar during 1982 and 1983. This was due to the loss of most small fish during 1982 but minimal loss of small fish after day 40 during 1983. The largest young-of-the-year reared during 1983 were quite a bit larger than the largest juveniles reared during 1982. Competition for food resulted in a wide size range during 1983. Condition of juvenile reared on *Tubifex* sp. and chopped earthworms was excellent.

Stocking

A total of 135 200, 1-day old embryos were stocked in the Red Cedar River in May of 1982. A total of 291 sturgeon weighing 8.2 kilograms were stocked in the Menominee River, in September. Of these fish, 250 were yearling sturgeon from the Center for Great Lakes Studies, University of Wisconsin-Milwaukee. Forty-one juveniles or yearlings were reared at Wild Rose State Hatchery during 1982.

In 1983, 82 000 8-day old embryos were stocked in the St. Louis River, in May; 1780 yearlings weighing 11 pounds were stocked there in August. 11 000 yearlings averaging 30 mm were stocked in the Menominee River in June.

Discussion and conclusion

Caesarian section proved to be the best way to collect eggs with the least harm to the adult fish. Incisions should be closed, using triangular shaped cutting needles. Round needles are inefficient when trying to penetrate the thick skin of sturgeon. Fish from which eggs have been taken by this method were radio tagged and post-surgical survival was found to be excellent (D.G. Folz, unpublished data) Milt may be held on ice for as long as 3 h and sperm still be viable. Eggs should be coated in bentonite clay to minimize clumping adhesion and thus decrease the probability of fungal infection. Eggs should be separated by pushing them through 4 mm mesh nylon. This may be done at any time during incubation without harm to the eggs.

A hatching temperature of 13° C was better than a temperature of 16° C. Results showed a higher percent of hatching success and more fully developed embryos when hatched in 13° C water. There were also fewer fungus problems.

Treatment of eggs with formalin (1:600) reduced, but did not eliminate problems with fungus. Astakhova & Martino (1968) observed inconsistent results using formalin concentrations of 0.05–0.5% to treat eggs. They also observed teratogenic effects which were not apparent in our study from treatment of eggs with formalin. They advocated the use of malachite green (1:200 000) and UV radiation to control fungus problems on eggs.

Lake sturgeon yearlings appear to prefer a diet of live food. Good growth and condition of yearlings were observed on diets of live brine shrimp nauplii, live *Tubifex* sp., live *Daphnia* sp., and chopped earthworms when fed in sufficient amounts. Poor survival and growth were observed when fish were fed dry diets, liver and fish mash. However, some fish were reared in previous years of the project on liver and dry feed. Frozen *Daphnia* sp. and frozen adult brine shrimp appeared to be marginal diets. Other researchers have also noted a marked preference for live food (Smith et al. 1980, Harkness & Dymond 1961).

The most critical factor in a lake sturgeon rearing program was to supply sufficient quantities of live food of increasing size while trying to gradually wean them to other diets. Large water volumes also appeared to be important. Water temperatures up to 21°C were found to increase growth. Frequent feedings of live food are recommended, as fish were observed to feed almost continuously when food was available. A wide selection of food items should be offered as soon as yolksacs are absorbed to minimize imprinting on one food item. This may help in weaning fish to a dry or frozen diet. Weaning from live feed should not be attempted until about 70 days when fish are growing rapidly.

Chloramine-T-hydrate appears to be an effective measure for disease prevention in yearling sturgeon. Sturgeon did not contract any diseases while being treated weekly during 1982. Concentrations of up to 16 milligrams per liter were safe in young more than 20 days old at temperatures from 13–22°C. Dip treatments with copper sulfate (1:2000) and salt (3%) (60 sec) were effective in controlling gill fungus during 1983.

References cited

Astakhova, T.V. & K.V. Martino. 1968. Measures for the control of fungous disease of the eggs of sturgeons in fish hatcheries. Journal of Ichthyology 8: 261–268.

Harkness, W.J.K. & J.R. Dymond. 1961. The lake sturgeon: the history of its fishery and problems of conservation. Ontario Dept. of Lands and Forests, Fish and Wildlife Branch, Maple. 121 pp.

Piper, R.G., I.B. McElwain, L.E. Orme, J.P. McCraren, L.G. Fowler & J.R. Leonard. 1982. Fish Hatchery management. U.S. Dept. of Interior, Fish & Wildlife Service, Washington D.C. 517 pp.

Smith, T.I.J., E.K. Dingley & D.E. Marchette. 1980. Induced spawning and culture of Atlantic sturgeon. Prog. Fish-Cult. 42: 147–150.

Received 15.5.1984   Accepted 14.2.1985

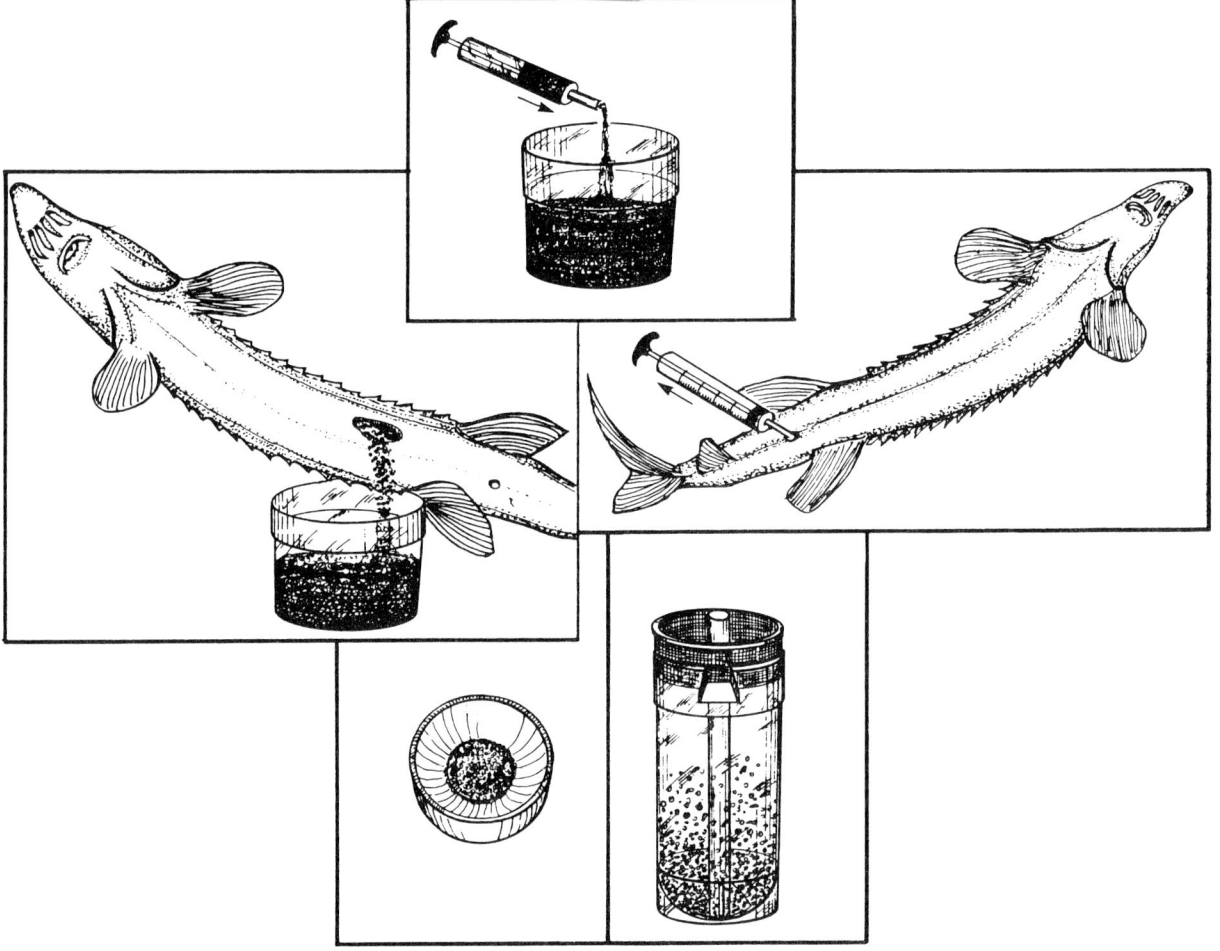

Sturgeon egg insemination and incubation includes the: surgical removal of ova, semen removal by catheterization, in vitro fertilization and egg incubation in a standard hatchery jar.

Oocyte maturation in white sturgeon, *Acipenser transmontanus*: some mechanisms and applications

Paul B. Lutes
Aquaculture Program, University of California, Davis, CA 95616, U.S.A.

Keywords: Chondrostean fish, Gonadotropins, Progestagens, Corticosteroids, Ovulation, Prostaglandins

Synopsis

The ovaries of four pre-spawning white sturgeon females were sampled and their oocytes incubated in the presence of eight gonadotropin preparations, 21 steroids, a prostaglandin and a catacholamine. Among the gonadotropin preparations, acetone dried pituitary gland powder from white sturgeon, common carp and chum salmon (in decreasing order of potency) were capable of inducing oocyte maturation (germinal vesicle breakdown – GVBD), while human chorionic gonadotropin, pregnant mare's serum gonadotropin, equine luteinizing hormone, bullfrog gonadotropin, and a stellate sturgeon pituitary chromatographic fraction capable of inducing testosterone production in white sturgeon testicular tissue failed to elicit any oocyte maturation response. The progesterone derivatives were the most potent steroid inducers of GVBD, followed closely by several corticosteroids. In vitro incubation of white sturgeon oocytes, in the presence of a suitable steroid (progesterone), can be used as a diagnostic tool in screening out unresponsive females for induced spawning work. The two remaining compounds, prostaglandin F_{2a} and epinephrine, failed to cause ovulation in progesterone-matured white sturgeon oocytes.

Introduction

This paper will focus on the internal environment of chemical signals controlling the final steps of oogenesis in the white sturgeon, *Acipenser transmontanus* Richardson, specifically those of final oocyte maturation and ovulation. It will include both basic data from several experiments describing the organismal, cellular and subcellular levels of chemical communication involved, and a discussion of the practical application of some of these data in the methodology of induced spawning.

Oocyte maturation is identified by the breakdown of the oocyte's nuclear membrane or germinal vesicle. Germinal vesicle breakdown (GVBD) is the indirect result of a surge in plasma gonadotropin. Specialized cells within the follicle recognize the increased amount of Gth, and in turn release a secondary messenger (Maturation Inducing Substance – MIS) which then directly interacts with the oocyte and ultimately induces GVBD.

Efforts to identify the MIS in ectothermic vertebrates began with studies on amphibian follicles (Schuetz 1974). It is generally accepted that among amphibians, progesterone acts as the in vivo MIS (Schorderet-Slatkine et al. 1980). The most common MIS reported among teleostean fishes, based on both in vitro and in vivo data, is 17α, 20β dihydroxy 4-pregnen-3-one (Nagahama et al. 1983). The chondrostean fishes have not been among the fish groups closely examined, but in vitro data, obtained for stellate sturgeon, *Acipenser stellatus*, has established that its oocytes are progesterone sensitive (Detlaf & Skoblina 1969, Skoblina et al. 1982).

In addition to oocyte maturation, the process of spawning in female sturgeon also includes ovulation, or the rupture of the follicular envelope and expulsion of the oocyte. Ovulation does not appear to be under the direct control of the MIS, but may instead be regulated by either a prostaglandin or a catacholamine, or both (Jalabert 1976, Jalabert & Szolosi 1975, Goetz et al. 1982, Goetz & Smith 1980, Stacey & Goetz 1982).

Materials and methods

Four white sturgeon females, ranging from 20 to 37 kg, were caught in San Francisco Bay prior to spawning migration and held at the hatchery as previously described (Doroshov et al. 1983, Doroshov & Lutes 1984); water temperatures during holding ranged from 12 to 15·C. Three to four months post-capture, the females' ovaries were preliminarily biopsied (Doroshov et al. 1983) to screen out non-responsive females. Any female with oocytes incapable of maturing in vitro when exposed to progesterone (10 μg ml^{-1}) and carp pituitary extract (100 μg ml^{-1}) were not used in subsequent incubations. Ovarian pieces from responsive females were then subdivided into groups of 20–30 follicles. These were then placed into plastic petri dishes containing 10 ml of Leibovitz' L-15 culture medium (GIBCO) to which penicillin (1000 I.U. l^{-1}), streptomycin (1000 μg l^{-1}) and fungizone (1.5 μg l^{-1}), an antimycotic, had been added. Upon completion of the incubation period (24 hours at 15°C, pH 7.2, without gassing), the oocytes were fixed in phosphate-buffered formalin (10% pH 7.4) for 48 hours, sectioned with a razor, and then scored for maturation by observing the frequency of germinal vesicle breakdown (GVBD).

All steroids used in the in vitro studies were obtained from Sigma Chemical Co., as were the human chorionic gonadotropin (HCG), the equine luteinizing hormone (LH), pregnant mares' serum gonadotropin (PMSG), prostaglandin F$_{2a}$ and epinephrine. The white sturgeon pituitary glands were from our own collection efforts, the common carp, *Cyprinus carpio*, pituitary material was obtained from Stoller, Inc., and the chum salmon, *Oncorhynchus keta*, pituitary material was a gift from Argent Chemicals. All pituitary material was from acetone dried glands. The sturgeon pituitary chromatographic fraction was originally derived from stellate sturgeon pituitaries collected in the USSR, and produced in the laboratory of Paul Licht (U.C. Berkeley). This fraction was capable of inducing testosterone synthesis in isolated white sturgeon testicular tissue (Licht, personal communication). Dr. Licht also donated the purified bullfrog gonadotropin. The steroid and Gth treatment ranges were based on teleostean studies (Kanatani & Nagahama 1981) and the preliminary screening of the females.

Steroids were administered after being dissolved in ethanol. The ethanol concentration in the culture medium was less than 1%. In the ovulation studies the prostaglandin F$_{2a}$ and epinephrine were added at four different times relative to the addition of progesterone (10 μg ml^{-1}): concurrent with the progesterone (0 h), and 10, 15, and 10 h post progesterone exposure.

Results

Among the gonadotropin treatments, only the three preparations from white sturgeon, common carp, and chum salmon pituitary glands were able to induce GVBD. None of the other compounds, including the sturgeon pituitary fraction, were observed to have any maturational effect on white sturgeon oocytes (0.001–1000 μg ml^{-1}). The lowest concentration of white sturgeon pituitary material to show any activity (70% GVBD) was 1.0 μg ml^{-1} of culture medium; 100% GVBD was observed at 100 μg ml^{-1}. The median effective dose (MED) was 0.56 μg ml^{-1} and had an MED of 35 μg ml^{-1}. Salmon pituitary material initially induced GVBD (15%) at 100 μg ml^{-1} and 100% GVBD was not observed at the dosages tested. Its MED was 340 μg ml^{-1}. All other substances (HCG, PMSG, equine LH, *A. stellatus* pituitary fraction and bullfrog Gth) were ineffective at the dosages tested.

Of the 21 steroid compounds tested, 17 exhibited at least a minimal ability to induce GVBD. 17α-hydroxy 4-pregnen-3-one, 17α, 20β dihydroxy 4-pregnen-3-one, and progesterone were the most active steroids, exerting their influence at concentrations as low as 3.9 ng ml^{-1} and evoking 100% GVBD at 62, 31, and 62 ng ml^{-1}, respectively (Fig. 1a). Several corticosteroids also exhibited significant maturational activity, although not as strongly as the progestagens. The most active corticoids included deoxycorticosterone (62 ng ml^{-1} 100% GVBD), cortisol (1.0 μg ml^{-1} 100% GVBD), and 11-deoxy cortisol (125 ng ml^{-1} 100% GVBD) (Fig. 1b). Minimally active steroids were androstenedione and aldosterone, neither of which were seen to induce 100% GVBD at the levels tested. Dehydroisoandrostenedione, estrone, estradiol, and estriol were all completely ineffective in inducing GVBD even at the highest concentrations (10 μg ml^{-1}) tested.

In the treatments for ovulation with 100% progesterone-induced maturation, neither prostaglandin F$_{2a}$ nor epinephrine caused ovulation at any of the concentrations tested (30 pg-10 μg ml^{-1}). The same results were obtained after delaying the introduction of either substance up to 20 hours after the initial exposure to progesterone.

Discussion

The results of the investigations presented here indicate possible patterns of endocrine control in follicular and oocyte maturation in the white sturgeon. White sturgeon oocytes respond in vitro to homoplastic pituitary material indicating a gonadotropin induced maturation, as established for teleostean fishes (de Vlaming 1983, Kanatani & Nagahama 1981, Hirose 1976, Yamaguchi & Yamamoto 1973). The gonadotropin target in white sturgeon is probably the follicular cells surrounding the oocyte, as established for *Acipenser stellatus* (Detlaff & Skoblina 1969). This is considered the general case among ectothermic vertebrates, although the Indian catfish, *Heteropneustus fossilis*, may be an exception (Sundararaj 1981). In *H. fossilis* the gonadotropin target appears to be the interrenal tissue, and the role of MIS is filled by a corticosteroid.

The potencies of common carp and salmon pituitary material did not approach that of the homoplastic material. The lack of response to LH, HCG, and PMSG is surprising, in light of their abilities to induce maturation in teleosts (Donaldson 1973), indicating the degree of separation in physiological evolution between chondrosteans and teleosts. Also worth noting was the inactivity of the stellate sturgeon pituitary fraction. This material, although observed to significantly increase testosterone production, which appears to control spermiation in teleosts (Hoar & Nagahama 1978), failed to elicit maturation in white sturgeon oocytes. This implies the presence of sex-specific isoforms of white sturgeon gonadotropin, as suggested by Idler et al. (1975), and Burlakov et al. (1979), or more than one Gth. The former hypothesis was tested by Goncharov et al. (1983) with four different isoforms of *A. stellatus* gonadotropin, chromatographically prepared on DAE-cellulose. Their studies, however, failed to confirm any sex-related differences among the variants tested.

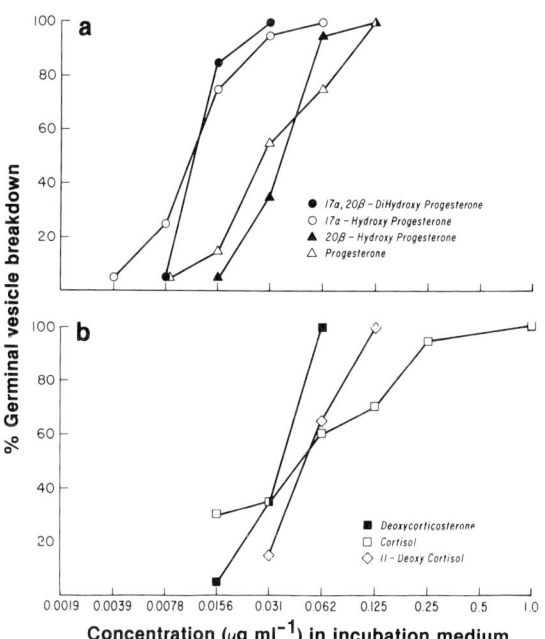

Fig. 1. Steroid dose-response curves for oocyte maturation – the seven most potent steroids: *a* – progestagens, *b* – corticosteroids.

The steroid incubations established the dominance of the progestagens as in vitro inducers of oocyte maturation in white sturgeon. Similar results have been reported in a variety of teleosts: biwamasu salmon (*Oncorhynchus masou rhodurus*), rainbow trout (*Salmo gairdneri*), and goldfish (*Carassius carassius*) (Nagahama et al. 1983), northern pike (*Esox lucius*) (Jalabert 1976), yellow perch (*Perca flavescens*) (Goetz & Theofan 1979), and medaka (*Oryzias latipes*) (Iwamatsu 1980). Furthermore, 17α, 20β dihydroxy 4-pregnen-3-one has been identified as being the most potent MIS among the progestagens in these fishes. Recent work by Nagahama et al. (1983) also shows a rise in plasma 17α, 20β dihydroxy 4-pregnen-3-one in *O. masou rhodurus* just prior to maturation. Conclusive identification of the specific MIS acting in white sturgeon awaits further studies incorporating dependably defolliculated oocytes, to avoid possible conversion of closely related precursors, and plasma steroid analysis during late vitellogenesis and meiotic maturation (Lutes et al. unpublished).

The positive maturational response to progestagenic steroids has an immediate practical application in the methodology of white sturgeon induced spawning. A simple in vitro oocyte incubation with an appropriate progestagen, e.g., progesterone (highly potent, but inexpensive), just prior to the initiation of induced spawning efforts, will effectively screen out a significant percentage of those females incapable of responding to exogenous hormonal treatment in the desired manner. The responsiveness of a minimum number of late stage oocytes biopsied from a given female can be determined after incubation under suitable conditions, simply by monitoring the rate of GVBD. This procedure was successfully utilized in the most recent white sturgeon spawning seasons (Lutes et al. unpublished).

In all gonadotropic and steroid incubations reported in this study, ovulation was seldom observed, never occurring in greater than 5% of the oocytes undergoing maturation. These results contrast with those of Detlaff & Skoblina (1969) in *A. stellatus* oocytes (involving both pituitary and progesterone incubations), wherein significant ovulation was reported. Dissociation of maturation and ovulation has been observed in teleosts (Jalabert 1976, Stacey & Goetz 1982, Espey 1980), and supports the presence of a separate endocrine mechanism for ovulation. In goldfish (Stacey 1976, Stacey & Peter 1979) a prostaglandin acts as the primary ovulatory agent, and preliminary evidence indicates that prostaglandins play a similar role in brook charr, *Salvelinus fontinalis* (Goetz 1980, Goetz & Smith 1980), yellow perch (Goetz & Theofan 1979), and rainbow trout (Jalabert 1976). White sturgeon oocytes did not respond to either prostaglandin F_{2a} or epinephrine, which induces ovulation in rainbow trout apparently via a-adrenergic receptors (Jalabert 1976), possibly by increasing general prostaglandin synthesis.

It must be noted here that in vitro white sturgeon oocyte maturation does not require an in vivo 'priming' of the contributing female with exogenous gonadotropin, as is the case with several teleost species, e.g. rainbow trout (Jalabert 1976), brook charr (Goetz & Bergman 1978), goldfish (Jalabert & Szollosi 1975), and mummichog *Fundulus heteroclitus* (Wallace & Selman 1978). In the yellow perch, however, no gonadotropin is required (Goetz & Theofan 1979) prior to maturation and ovulation induction. Yellow perch and white sturgeon therefore present 'cleaner' systems for examining the mechanisms of ovulation.

The incidence of maturation without ovulation in vivo is a phenomenon sometimes observed in the practical induced spawning of several cultured fish species. Such female fish exhibit many of the signs associated with spawning readiness – swollen abdomen, red and swollen vent, specific behavioral patterns – but fail to spawn properly. The terms commonly use for such fish are 'plugged', 'overripe', and 'egg-bound', and this response has been encountered in attempts to spawn the white sturgeon (Doroshov et al. 1983). Upon examination of the white sturgeon oviducts, however, no obstructions are observed and the follicles within the ovary are seen to still contain their oocytes, although they have undergone GVBD. This may simply reflect an abnormal state of oogenic development, in that the follicle is incapable of responding to the ovulatory signal. The occurrence of a physiological mechanism bypassing the ovulatory stimulus and perhaps leading into atresia, however, must also be considered.

In this study preliminary evidence indicating a possible role for 17α, 20β dihydroxy 4-pregnen-3-one as the MIS in the white sturgeon has been presented. This is in agreement with the data available for teleostean fishes. Unlike several teleosts examined to date, however, white sturgeon do not exhibit an ovulatory response to either prostaglandin F_{2a} or epinephrine.

References cited

Burlakov, A.B., A.P. Zolotnitskii & E.B. Moiseeva. 1979. Pituitary gonadotropic hormone from the Black Sea plaice Scopthalmus maeoticus. Zh. Evol. Biokhim. Fiziol. 15: 496–499.

Donaldson, E.M. 1973. Reproductive endocrinology of fishes. Amer. Zool. 13: 909–927.

Detlaff, T.A. & M.N. Skoblina. 1969. The role of the germinal vesicle in the process of oocyte maturation in Anura and Acipenseridae. Annales D'Embryol. Morphol Suppl. 1: 133–151.

Doroshov, S.I., W.H. Clark, Jr., P.B. Lutes, R.L. Swallow, K.E. Beer, A.B. McGuire & M.D. Cochran. 1983. Artificial propagation of the white sturgeon Acipenser transmontanus. Aquaculture 32: 93–104.

Doroshov, S.I. & P.B. Lutes. 1984. Preliminary data on the induction of ovulation in the white sturgeon (Acipenser transmontanus). Aquaculture 38: 221–227.

Espey, L.L. 1980. Ovulation as an inflammatory response – a hypothesis. Biol. Reprod. 22: 73–106.

Goetz, F.W. 1980. Prostaglandin levels during steroid-induced in vitro final maturation of brook trout (Salvelinus fontinalis) oocytes. Amer. Zool. 20: 730.

Goetz, F.W. & H.L. Bergman. 1978. The effects of steroids on final maturation and ovulation of oocytes from brook trout (Salvelinus fontinalis) and yellow perch (Perca flavescens). Biol. Reprod. 18: 293–298.

Goetz, F.W. & D.C. Smith. 1980. In vitro effects of theophylline and dibutyryl adenosine 3':5' cyclic monophosphoric acid on spontaneous and prostaglandin F_{2a} – induced ovulation of brook trout (Salvelinus fontinalis) oocytes. Biol. Reprod. 22: 114A.

Goetz, F.W., D.C. Smith & S.P. Krickle. 1982. The effects of prostaglandins, phosphodiesterase inhibitors, and cAMP on ovulation of brook trout (Salvelinus fontinalis) oocytes. Gen. Comp. Endocrinol. 48: 154–160.

Goetz, F.W. & G. Theofan. 1979. In vitro stimulation of germinal vesicle breakdown and ovulation of yellow perch (Perca flavescens) oocytes: effects of 17α, 20β dihydroxy progesterone and prostaglandins. Gen. Comp. Endocrinol. 37: 273–385.

Goncharov, B.F., A.A. Kuznetsov & E. Burzawa-Gerard. 1983. Pituitary gonadotropic hormone from a chondrostean fish, the starred sturgeon (Acipenser stellatus), IV. Differences in biological action of isoforms and absence of sex-specific forms. Gen. Comp. Endocrinol. 49: 375–382.

Hirose, K. 1976. Endocrine control of ovulation in medaka (Oryzias latipes) and ayu (Plecoglossus altivelis). J. Fish Res. Board Can. 33: 989–994.

Hoar, W.S. & Y. Nagahama. 1978. The cellular sources of sex steroids in teleost gonads. Ann. Biol. Anim. Biochim. Biophys. 18: 893–898.

Idler, D.R., L.S. Bazar & S.L. Hwang. 1975. Fish gonadotropin(s) III. Evidence for more than one gonadotropin chum salmon pituitary glands. Endocrinol. Res. Commun. 2: 237.

Iwamatsu, T. 1980. Studies in oocyte maturation of the medaka (Oryzias latipes) VIII. Role of follicular constituents in gonadotropin- and steroid-induced maturation of oocytes in vitro. J. Exp. Zool. 211: 231–239.

Jalabert, B. 1976. In vitro oocyte maturation and ovulation in rainbow trout (Salmo gairdneri), northern pike (Esox lucius), and goldfish (Carassius carassius). J. Fish. Res. Board Can. 33: 974–988.

Jalabert, B & D. Szollosi. 1975. In vitro ovulation of trout oocytes: Effects of prostaglandins in smooth muscle-like cells of the theca. Prostaglandins 9: 765–768.

Kanatani, H. & Y. Nagahama. 1981. Mediators of oocyte maturation. Biomed. Res. 1: 273–291.

Nagahama, Y, K. Hirose, G. Young, S. Adachi, K. Suzuki & B. Tamaoki. 1983. Relative in vitro effectiveness of 17α 20β dihydroxy 4-pregnen-3-one and other pregnene derivatives on germinal vesicle breakdown in the oocytes of ayu (Plecoglossus altivelis), amago salmon (Onchorynchus rhodurus), rainbow trout (Salmo gairdneri) and goldfish (Carassius carassius). Gen. Comp. Endocrinol. 51: 15–23.

Schorderet-Slatkine, S., M. Schorderet, M. El-Etr, F. Godeau, P. Boquet & E. Baulieu. 1980. The role of cAMP in the process of meiosis reinitiation in Xenopus laevis oocytes: Effects of progesterone, insulin, and cholera toxin. pp. 93–101. In: G. Delrio & J. Brachet (ed.) Steroids and their Mechanisms of Action, Raven Press, New York.

Schuetz, A.W. 1974. The role of hormones in oocyte maturation. Biol. Reprod. 10: 150–178.

Skoblina, M.N., Zh.G. Shmerling & O.T. Kondratieva. 1981. Cholesterol-induced in vitro maturation of oocytes of Acipenser stellatus, Xenopus laevis, and Rana temporaria. Gen. Comp. Endocrinol. 44: 470–475.

Stacey, N.E. 1976. Effects of indomethacin and prostaglandins on spawning behavior of female goldfish. Prostaglandins 12: 113–126.

Stacey, N.E. & F.W. Goetz. 1982. Role of prostaglandins in fish reproduction. Can. J. Fish. Aquat. Sci. 39: 92–98.

Stacey, N.E. & S. Pandey. 1975. Effects of indomethacin and prostaglandins on ovulation in goldfish. Prostaglandins 9: 597–608.

Stacey, N.E. & R.E. Peter. 1979. Central action of prostaglandins in spawning behavior of female goldfish. Physiol. Behav. 22: 1191–1196.

Sundararaj, B.I. 1981. Reproductive physiology of teleost fishes. Food and Agricultural Organization of the United Nations, Rome, ADCEP/REP/81/16. 82 pp.

de Vlaming, V. 1983. Oocyte development patterns and hormonal involvements among teleosts. pp. 176–199. In: J.C. Rankin, T.J. Pitcher & R. Duggen (ed.) Controlled Processes in Fish Physiology, Croomhelm, London.

de Vlaming, V. 1972. Environmental control of teleost reproductive cycles: a brief review. J. Fish Biol. 4: 131–140.

Wallace, R.A. & K. Selman. 1978. Oogenesis in *fundulus heteroclitus* I. Preliminary observations on oocyte maturation *in vivo* and *in vitro*. Devel. Biol. 62: 354–359.

Yamaguchi, K. & Yamamoto. 1973. *In vitro* maturation of the oocytes in the medaka (*Oryzias latipes*). Annot. Zool. Jap. 46: 144–153.

Reveived 15.5.1984   Accepted 12.3.1985

The effect of food deprivation on the plasma free amino acid levels of sturgeon, *Acipenser transmontanus*

Richard L. Swallow
Coker College, Hartsville, SC 29550, U.S.A.

Keywords: Leucine, Isoleucine, Valine, Herring roe, Peripheral blood, Liver, Feeding, Tyrosine, Asparagine, Catabolism, Chondrostean fish

Synopsis

Plasma levels of free amino acids were compared in recently fed sturgeon and in those deprived of food for over two weeks. Fed animals had elevated levels of branched chain amino acids. It is suggested that the higher levels of these amino acids are due to reduced metabolism of these amino acids as well as high levels in the food.

Introduction

The relationships between the free amino acid composition of diet and plasma levels of free amino acids in fish have been studied by many workers recently. Nose (1972) found that essential amino acid levels in the plasma correlated well with their proportion in the diet in rainbow trout, *Salmo gairdneri*. Plakas et al. (1980) found similar results in carp, *Cyprinus carpio*. In a later study on carp, Debrowski (1982) showed similar results even when distributions between erythrocytes and plasma were taken into consideration. However, Kaushik & Luquet (1979) found such correlations in aromatic, sulfur and branched chain amino acids, but not in the others. They also pointed out that correlations between diet and plasma levels of amino acids should not necessarily be expected in view of the fact that plasma levels of animo acids are the result of efflux from many compartments, not just from diet.

All of these previous studies were carried out on fish held for relatively short periods without food on fish fed and held under laboratory conditions. This report contains data on plasma free amino acid levels on sturgeon fed recently in the wild and on sturgeon held for 14 to 18 days after feeding on the same diet.

Materials and methods

Female white sturgeon, *Acipenser transmontanus*, were collected from San Francisco Bay. They were slowly acclimated to fresh water either at the Tiburon National Marine Laboratory or after transfer to the Davis campus. The animals had been feeding upon herring, *Clupea pallasi*, roe at the time of capture. Herring roe samples were taken from the guts of three sturgeon after capture.

One group of fish had blood samples removed immediately upon capture. These animals had intestine full of herring roe. A second group of fish were transported to the Davis campus. These fish were maintained in large holding tanks (2.8 × 6 m) supplied with a continuous flow of fresh well water. The temperature of the holding tanks varied from 14 to 18° C depending upon the well water temperature. These animals had blood samples removed 14 to 18 days after capture.

Blood samples of 10 to 12 ml were taken by direct heart puncture from each fish. After the blood was drawn, it was quickly cooled on ice, plasma separated from cells, frozen on dry ice and held for later analysis.

For amino acid analysis plasma was deproteinized with sulfosalicyclic acid, and the pH was adjusted to 2.0 ± 0.1. Herring roe samples were subjected to acid hydrolysis with 6N HCl in sealed tubes. All samples were analyzed on a Beckman 121 Amino Acid Analyzer.

Differences in plasma levels of amino acids were statistically analyzed using a Students t-test.

Results

The effects of food deprivation on the levels of free amino acids in the plasma of sturgeon are shown in Table 1. The concentrations of all amino acids decreased over 25% during the period of food deprivation. Most of the decrease in total free amino acids was due to changes in three amino acids — valine, leucine, and isoleucine. Tyrosine levels decreased slightly while asparagine was the only amino acid to show a significant increase during food deprivation.

Food deprivation also reduced the relative concentrations of the three branched-chain amino acids (Table 2). Leucine, isoleucine and valine contributed 22.78%, 9.62%, and 22.05% respectively to the plasma levels of amino acids in fed animals. These amino acids represented over 53% of the total amino acids present in fed animals. Even in the food deprived animals these three amino acids contributed over 31% of the total amino acids present. Table 2 also contains the relative proportions of amino acids contained in the herring roe upon which the sturgeon were feeding. The increases in the proportions of the branched-chain amino acids in the plasma of fed animals appear to correlate well with their relative proportions in the roe. The other amino acid plasma levels do not appear to be affected by the levels in the herring roe.

Table 1. Comparison of levels of free amino acids in the plasma of food deprived and fed female sturgeon (uM L^{-1}).

Amino acid	Food deprived (N = 23)	Fed (N = 14)
Arg	105 ± 12	74 ± 15
His	42 ± 8	33 ± 15
Ile	157 ± 13	328 ± 31*
Leu	302 ± 27	777 ± 73*
Lys	303 ± 25	268 ± 37
Met	46 ± 6	44 ± 13
Phe	74 ± 11	109 ± 17
Thr	181 ± 22	138 ± 33
Try	171 ± 26	169 ± 48
Val	301 ± 25	752 ± 66*
Ala	293 ± 37	255 ± 51
Asn	41 ± 10	12 ± 4*
Asp	12 ± 1	13 ± 5
Cys	9 ± 5	14 ± 4
Glu	14 ± 5	12 ± 2
Gln	851 ± 120	763 ± 138
Gly	25 ± 6	17 ± 2
Ser	93 ± 7	63 ± 41
Pro	133 ± 30	221 ± 40
Tyr	58 ± 6	124 ± 22*

Values represent mean ± S.E., * P<0.05 comparing food deprived and fed.

Table 2. Comparison of percentage contribution of individual amino acids to plasma and herring roe.

Amino acid	Food deprived	Fed plasma	Herring roe
Arg	4.53	2.17	3.30
His	1.81	0.97	2.00
Ile	6.77	9.62	6.80
Leu	13.02	22.78	11.50
Lys	13.07	7.86	6.80
Met	1.98	1.29	0.80
Phe	3.19	3.20	2.80
Thr	7.81	4.05	6.30
Try	7.37	4.95	–
Val	12.98	22.05	8.20
Ala	12.63	7.48	13.90
Asp	0.52	0.38	7.50
Glu	0.60	0.35	10.10
Gly	1.08	0.50	5.90
Ser	4.01	1.85	5.50
Pro	5.74	6.48	6.20
Tyr	2.50	3.64	2.40

Discussion

The data presented in this study indicate that the levels of the branched-chain amino acids decrease upon food deprivation, and that the higher level in the amino acids in recently fed sturgeon may be at least due in part to the high relative levels of these amino acids in the food. Similar increases were found in essential amino acids in rainbow trout (Nose 1972, Schlesio & Nicolai 1978) and in carp (Plakas et al. 1980, Dabrowski 1982). The essential amino acids have not been established for the sturgeon, but it is interesting to note that no other amino acids in the essential group of other fishes were elevated in recently fed sturgeon while other essential amino acids did increase in trout and carp.

As Dabrowski (1982) has suggested, the removal of amino acids from the plasma after feeding could be the result of catabolism, protein synthesis or tissue storage. But it is also important to distinguish between those processes which remove amino acids from the plasma compartment and those which contribute amino acids to the plasma. This report and the others cited above all present amino acid patterns in peripheral blood. No data is available concerning levels in portal blood of fishes versus levels in peripheral blood in the period immediately following feeding. Thus, all reports on fish view the results of liver metabolism on the absorbed amino acids from the food.

There are reports of the effect of the liver on plasma levels of amino acids in peripheral blood after feeding in mammals. Elwyn et al. (1968) working with dogs showed that the liver acts to buffer the peripheral circulation from fluctuations in individual amino acid concentrations except in the case of the branched-chain amino acids. It is very likely that sturgeon and other fishes show elevated levels of the branched-chain amino acids as well as other essential amino acids because the liver is unable to remove them efficiently from the blood.

References cited

Dabrowski, K. 1982. Postprandial distribution of free amino acids between plasma and erythrocytes of common carp (*Cyprinus carpio* L.). Comp. Biochem. Physiol. 72A: 753–763.

Elwyn, D.H., H.C. Parikh & W.C. Shoemaker. 1968. Amino acid movements between gut, liver and periphery in unanesthetized dogs. Amer. J. Physiol. 215: 1260–1275.

Kaushik, S.J. & P. Luquet. 1979. Influence of dietary amino patterns on the free amino acid contents of blood and muscle of rainbow trout (*Salmo gairdnerii* R.). Comp. Biochem. Physiol. 64B: 175–180.

Nose, T. 1972. Changes in the pattern of free plasma amino acids in rainbow trout after feeding. Bull. Freshw. Fish. Res. Lab. 22: 137–144.

Plakas, S.M., T. Katayama, Y. Tanaka & O. Deshimaru. 1980. Changes in the levels of circulating plasma free amino acids of carp (*Cyprinus carpio*) after feeding a protein and an amino acid diet of similar composition. Aquaculture 21: 307–322.

Schlisio, W. & B. Nicolai. 1978. Kinetic investigations on the behaviour of free amino acids in the plasma and of tow aminotransferases in the liver of rainbow trout (*Salmo gairdneri* R.) after feeding on a synthetic compostion containing pure amino acids. Comp. Biochem. Physiol. 59B: 373–379.

Received 15.5.1984   Accepted 10.3.1985

Sturgeon eggs clearly demonstrate holoblastic cleavage, hatching embryos have underdeveloped major organs (mouth, fins, digestive system), external feeding begins about 10 to 15 days after hatching.

Evaluation of morphometric characters used in taxonomic separation of Gulf of Mexico sturgeon, *Acipenser oxyrhynchus desotoi*

Charles M. Wooley
U. S. Fish and Wildlife Service, Office of Fishery Assistance, 1612 June Avenue, Panama City, FL 32405, U.S.A. Current address: USFWS c/o Fisheries Division, Tidewater Administration, Maryland DNR, C-2 Tawes Building, Annapolis, MD 21401, U.S.A.

Keywords: Atlantic sturgeon, Subspecies, Taxonomy, Morphometrics, *Chondrosteans*, Management, Stocks, Anadromous, Spleen, Apalachicola River, Florida

Synopsis

Morphometric variation within *Acipenser oxyrhynchus desotoi* is summarized, and the utility of characters that have been used to recognize the subspecies are evaluated. Data derived from measurements of two juvenile specimens used by Vladykov (1955) to define the currently described subspecies *A. o. desotoi* provide the standard for comparison. Head size, dorsal scute shape, and pectoral fin length indicate statistically significant (t-tests; $p = 0.05$) differences between *A. o. desotoi* and *A. o. oxyrhynchus*, although there is a certain amount of value overlap for each characteristic. The spleen length versus fork length ratio is the only morphometric characteristic that has only a small degree of value overlap and approaches being totally diagnostic. The two subspecies are allopatric, with the range of *A. o. desotoi* being the northeastern Gulf of Mexico from Tampa Bay, Florida, westward to at least the Mississippi River. Pending a comprehensive analysis of the *A. oxyrhynchus* complex throughout its range, *A. o. desotoi* is tentatively retained as a valid subspecies.

Introduction

The effective management of sturgeon in the Gulf of Mexico requires a clear understanding of *A. o. oxyrhynchus* subspecies composition and stocks. However, for the last 10 years, the distinction between Atlantic sturgeon (*A. o. oxyrhynchus*) and the subspecies Gulf of Mexico sturgeon (*A. o. desotoi*) has not been clear. Little information regarding morphometrics, meristics and ecology of the subspecies *A. o. desotoi* is available. This paper presents morphometrics of sturgeon captured in conjunction with tagging and radio telemetry studies. The study objectives being: (1) to determine if Gulf of Mexico sturgeon from the Apalachicola River, and other Gulf of Mexico drainages in Florida, could be discriminated into the subspecies *A. o. desotoi* on the basis of four morphological characteristics as proposed by Vladykov (1955), and (2) to compare our morphological measurements with Vladykov's holotype and paratype specimens, and additional *A. o. oxyrhynchus* specimens from the Atlantic ocean.

Range

The Atlantic sturgeon, *A. o. oxyrhynchus*, ranges from Hamilton Inlet, Labrador, Canada, south at least to the St. Johns River, Florida (Gilbert 1978, Wooley & Crateau 1982, 1985). Cold fronts along the Atlantic seaboard induce a few *A. o. oxyrhynchus* to move southward along the eastern Florida coast during the winter to as far south as Port Canaveral (UF/FSM 30903) and Hutchinson Island, Florida (Grant Gilmore, Harbor Branch Foundation, personal communication).

The Gulf of Mexico sturgeon, *A. o. desotoi*, occurs in the northern Gulf of Mexico where it historically ranged from Tampa Bay, Florida, west to at least the mouth of the Mississippi River (Vladykov & Greeley 1963). Two specimens have been recorded from Charlotte Harbor, Florida (UF/FSM 35332, and a mounted specimen at the Florida Department of Natural Resources laboratory, St. Petersburg, Florida), but these are exceptions to its present range in the Gulf of Mexico which appears to be from the Suwannee River area in west-central Florida (Huff 1975) to eastern Louisiana (Davis et al. 1970). These subspecies are allopatric, having been separated by the emergence of peninsular Florida and maintained by the thermal barrier of the Gulf Stream around south Florida (Rivas 1954).

Methods

From 1981–1983 sturgeon were collected from the Apalachicola River between river kilometer 168.9 and 170.5, with sinking monofilament and multifilament (114–305 mm stretch mesh) gill nets and by electrofisher. Measurements and proportional data were taken according to Bailey & Cross (1954), Vladykov (1955) and Hubbs & Lagler (1958). Measurements on specimens were performed with dividers and calipers; all measurements were taken in a straight line. Because of the tenuous status of the *A. o. desotoi* population, we did not plan to sacrifice any fish for internal measurements. However, we were able to examine the internal organs of three *A. o. desotoi* due to accidental mortality, and nine museum specimens.

Material examined

A. o. desotoi: Eighty-seven *A. o. desotoi* were examined during tagging operations on the Apalachicola River (Wooley & Crateau 1985) one from the Choctawhatchee River and one specimen from the Escambia River, all in Florida. Nine *A. o. desotoi* specimens (Table 1) were examined from the Florida State Museum (UF/FSM), University of Florida, Gainesville. *A. o. oxyrhynchus*: Sixteen *A. o. oxyrhynchus* specimens were examined from the Florida State Museum and eight from American Museum of Natural History (AMNH), New York. (Tables 2, 3).

Characters

Specimens were examined for (1) head length over fork length; (2) left pectoral fin length over fork length; (3) length over width of the fifth dorsal scute; (4) length over width of the sixth dorsal scute. On 35 dead sturgeon internal morphometric characters were examined: (1) spleen length versus fork length; (2) spleen length, expressed in percentage of distance between the lower end of the pyloric apparatus and lower end of the intestinal loop; (3) pyloric apparatus length over fork length; (4) spleen length over the length of pyloric apparatus.

Table 1. Acipenser oxyrhynchus desotoi examined from the University of Florida, Florida State Museum (UF/FSM) collection.

UF/FSM identification	Number examined	Fork length mm	Collection location
28574	1	445	Suwannee River, Florida
35332	1	1070	Charlotte Harbor, Florida
35373	1	520	Choctawhatchee Bay, Florida
24610	1	381	Suwannee River, Florida
24611	1	392	Suwannee River, Florida
24612	1	445	Suwannee River, Florida
37723	1	556	Apalachicola River, Florida
37724	1	503	Apalachicola River, Florida
37725	1	445	Apalachicola River, Florida

Table 2. Acipenser oxyrhynchus oxyrhynchus examined from the University of Florida, Florida State Museum (UF/FSM) collection.

UF/FSM identification	Number examined	Fork length mm	Collection location
7323	1	203	Connecticut River, Massachusetts
25761	7	339–453	Hudson River, New York
30903	1	580	Port Canaveral, Florida
30904	1	482	Altamaha River, Florida
30956	1	660	St. Johns River, Florida
35719	1	417	Cape Fear River, North Carolina
2379	4	165–285	St. Lawrence River, Quebec

Table 3. Acipenser oxyrhynchus oxyrhynchus examined from the American Museum of Natural History (AMNH) collection.

AMNH identification	Number examined	Fork length mm	Collection location
48339	1	69	Hudson River, New York
39272 (A)	1	451	Hudson River, New York
39272 (B)	1	434	Hudson River, New York
39272 (C)	1	425	Hudson River, New York
39272 (D)	1	452	Hudson River, New York
39272 (E)	1	299	Hudson River, New York
39272 (F)	1	460	Hudson River, New York
39272 (G)	1	1904	Hudson River, New York

Results and discussions

Vladykov (1955) stated that the two subspecies, *A. o. oxyrhynchus* and *A. o. desotoi* are easily distinguished by body proportions, differences in the shape and degree of development of the dorsal scutes, as well as in the details of interior anatomy.

According to Vladykov (1955) the subspecies *A. o. desotoi* is distinguishable from *A. o. oxyrhynchus* by a large head (30.9–33.6% of FL) and longer pectoral fins (15.6–16.3% of FL) and dorsal scutes the length of which is much shorter than the width. The spleen in *A. o. desotoi* is unusually long, extending from the pybus to the lower end of the loop of the small intestine, while in *A. o. oxyrhynchus* this organ does not reach below the middle of the loop.

The type and paratype specimens (Field Museum of Natural History, Chicago 59803 and 59804) were collected on November 30, 1953 by T. Dawson from between Twin and Rabbit Islands at the mouth of the Singing River, off Gautier, Mississippi, U.S.A. Vladykov (1955) provided a detailed description of the holotype and paratype. Both fish were sexually immature with fork lengths of 515 and 595 mm. This location is 370 km west of the Apalachicola River, with the Apalachicola River the approximate mid-point of the *A. o. desotoi* range.

Body proportions

Head length-Fork length — Ninety-eight *A. o. desotoi*, ranging in fork length from 360–2110 mm, have been observed over the last 3 years with a head length over fork length mean percent measurement of 29.4% (range 27.0–33.3%, S.D. 1.4). There is a highly significant relationship between increased fork length and decreasing head length ($P<0.001$; $n = 98$, $r = -0.801$). In 39 specimens of *A. o. desotoi* ranging in length from 600–900 mm fork length, the head was proportionately larger (mean of 29.6%) than in the 48 specimens above 900 mm (mean of 28.6%). In small specimens of *A. o. desotoi* ranging in length from 360–599 mm fork length the head length over fork length ratio was proportionately larger (32.6%).

The two head length versus fork length body proportion measurements obtained by Vladykov (1955) for his type and paratype specimens were 30.9 and 33.9%. Our mean of 29.4% for *A. o. desotoi* is less than the 32.3% recorded by Vladykov (1955) for *A. o. desotoi*, although Vladykov's values fall within these ranges.

Head length versus fork length measurements for 22 museum specimens of *A. o. oxyrhynchus* (69–1904 mm fork length) I examined averaged 27.4%. These specimens had a range of 24.2–33.8% (S.D. 3.0), had a highly significant relationship between increased fork length and decreasing head length (P <0.001; df = 20; r = −0.673), and agree with Vladykov's (1955) mean of 27.0% for *A. o. oxyrhynchus* (Fig. 1, 2).

As previously discussed, the head length versus fork length percentage has been traditionally used to distinguish *A. o. desotoi* from *A. o. oxyrhynchus*. My values of head length versus fork length mean of 29.4% for *A. o. desotoi* indicate a statistically significant difference (t – tests; P = 0.05) between the head length versus fork length measurement mean of 27.4% for *A. o. oxyrhynchus*. However, in small (300–600 mm fork length) specimens of *A. o. desotoi*, the head is proportionately larger than in specimens greater than 900 mm fork length. As the fork length of *A. o. desotoi* increases, the head length over fork length percentage decreases due to allometric growth. Because of allometric growth the head length versus fork length measurement is too variable a characteristic to be useful in subspecies identification. This morphological feature is not diagnostic for *A. o. desotoi* over 1200 mm fork length, which average a head length versus fork length measurement of <28%, the same percentage range which was observed from museum specimens of *A. o. oxyrhynchus* sturgeon between 300–600 mm fork length. It is interesting to note though, that in specimens of proportional fork length, the grouping of head length versus fork length data for each of the subspecies forms does not overlap.

Pectoral fin length—Fork length — Mean left pectoral fin length – fork length for 23 *A. o. oxyrhynchus* was 13.6% (range 12.0–15.7%, S.D.

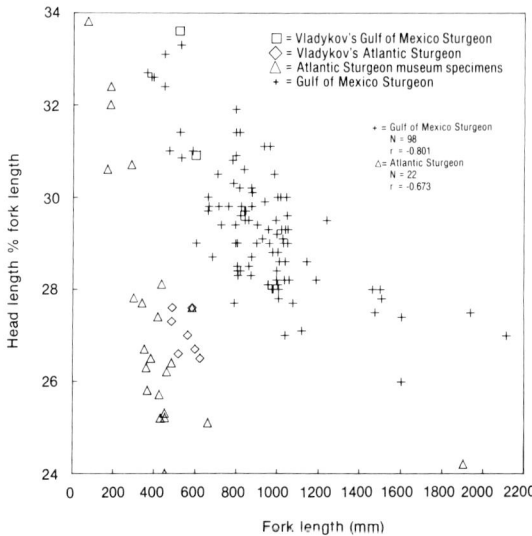

Fig. 1. Scatter plot depicting proportional length of the head in various size groups and collections of *Acipenser oxyrhynchus desotoi* and *Acipenser oxyrhynchus oxyrhynchus*. Regression equations, line for *A. o. desotoi* (Y = 31.891X – 0.002); line for *A. o. oxyrhynchus* (Y = 29.772X – 0.005).

0.80). The mean of 13.6% agrees closely with Vladykov's mean of 13.4% for *A. o. oxyrhynchus* and our ranges show little overlap. Vladykov's (1955) observed mean of 15.9% for *A. o. desotoi* is approximately 3.5 percentage points greater than our observed mean of 12.5% (range 10.4–15.9%, S.D. 0.74). The upper range of values does agree with Vladykov's mean, but the sample size of 84 individuals shows a greater range of values, and a normal distribution of values, when compared to Vladykov's sample of only two individuals.

Results indicate a statistically significant difference in pectoral fin length in *A. o. desotoi* (P = 0.05) than in *A. o. oxyrhynchus*. *A. o. desotoi* has a smaller average pectoral fin length versus fork length ratio than *A. o. oxyrhynchus*, an exact opposite characteristic than that observed by Vladykov (1955).

Dorsal scutes — Average length versus width of the fifth dorsal scute in 48 *A. o. desotoi* was 0.89% (range 0.65–1.27%, S.D. 0.14). There was no significant relationship between the dorsal scute length versus width measurement and increased fork length (r = 0.003). Average length versus

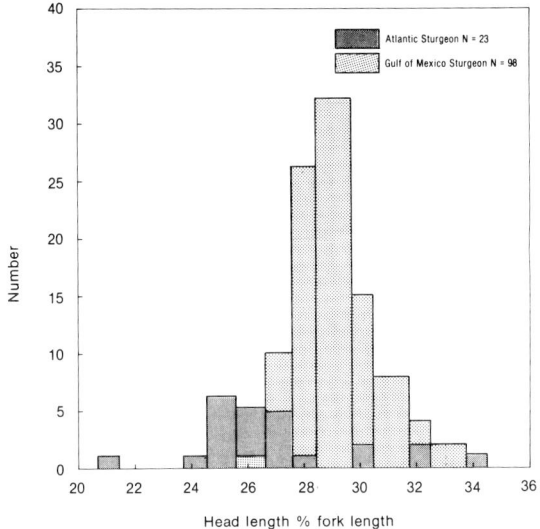

Fig. 2. Frequency and range of head length in percent of fork length in *Acipenser oxyrhynchus desotoi* and *Acipenser oxyrhynchus oxyrhynchus*.

width of the sixth dorsal scute in 51 *A. o. desotoi* was 0.87% (range 0.70–1.21%, S.D. 0.11) with no significant relationship between the length versus width (r = 0.217). The means concur with Vladykov's (1955) statement that in *A. o. desotoi*, dorsal scutes are more square, with the length being shorter than the width than in *A. o. oxyrhynchus*.

Average length versus width of the fifth dorsal scute in 23 *A. o. oxyrhynchus* was 1.11% (range 0.83–1.82%, S.D. 0.25), and for the sixth dorsal scute 1.16% (n = 23, range 0.83 – 1.67%, S.D. 0.20). There is a statistically significant difference (P = 0.05) between *A. o. desotoi* and *A. o. oxyrhynchus* in the fifth and sixth dorsal scute measurements, despite the overlap in ranges between the two subspecies (Table 4).

Interior anatomy

Spleen — Vladykov (1955) states that the most striking characteristic of the viscera in *A. o. desotoi* is the length of the spleen. The length of the spleen of the two specimens Vladykov (1955) examined, expressed in percentage of the fork length was 16.0 and 19.0%. The length of the spleen versus fork length for 13 specimens of *A. o. oxyrhynchus* examined by Vladykov ranged from 3.0–9.0%.

Twelve *A. o. desotoi* specimens were examined and indicated a mean spleen length versus fork length measurement of 12.3 (range 7.9–15.8%, S.D. 2.5, r = 0.212). Twenty two *A. o. oxyrhynchus* were examined which had a mean spleen length versus fork length measurement of 5.7% (range 2.8–8.3%, S.D. 1.8, r = 0.121). There is a statistically significant difference (P = 0.05) between the spleen length versus fork length measurements obtained in *A. o. desotoi* and *A. o. oxyrhynchus* samples with minimal overlap between the minimum and maximum values (Fig. 3).

The length of the spleen expressed in percentage of distance between the lower end of the pyloric apparatus and lower end of the small intestine loop according to Vladykov (1955) is 80–122% (average 101.0%) in *A. o. desotoi*, and 19.0–42.0% (average 31.0%) in *A. o. oxyrhynchus*. Twelve *A. o. desotoi* examined averaged 76.8% (range 56.8–109.3%, S.D. 15.5, r = 0.343) in distance between the lower

Table 4. Body proportions of *Acipenser oxyrhynchus oxyrhynchus* and *Acipenser oxyrhynchus desotoi*.

Body proportions	Subspecies Locality	*A. o. desotoi* Gulf of Mexico Mean (N)	*A. o. desotoi*[1] Gulf of Mexico Mean (N)	*A. o. oxyrhynchus*[1] Atlantic Ocean Mean (N)	*A. o. oxyrhynchus* Atlantic Ocean Mean (N)
Fork length (mm)		830 (98)	555 (2)	559 (6)	440.3 (24)
Head length in % fork length		29.4 (98)	32.3 (2)	27.0 (6)	27.4 (22)
Left pectoral in % fork length		12.5 (84)	15.9 (2)	13.4 (6)	13.6 (23)
5th scutes length in % width		0.89 (48)	–	–	1.1 (23)
6th scutes length in % width		0.87 (51)	0.81 (2)	1.13 (3)	1.2 (23)

[1] Vladykov (1955)

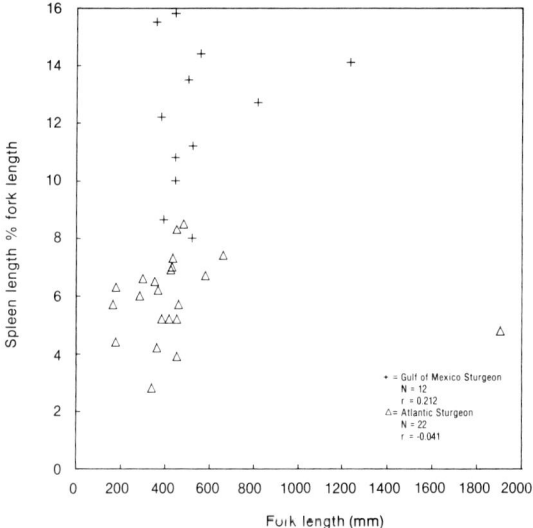

Fig. 3. Spleen length versus fork length measurements of *Acipenser oxyrhynchus desotoi* and *Acipenser oxyrhynchus oxyrhynchus*. Regression equations, line for *A. o. desotoi* (Y = 303.632X + 20.269); line for *A. o. oxyrhynchus* (Y = 524.234X − 11.261).

end of the pyloric apparatus and lower end of the small intestine loop. Twenty three *A. o. oxyrhynchus* averaged 39.8% (range 21.1–70.5%, S.D. 12.0) for the same measurement with overlap between the minimum and maximum values (Table 5).

The pyloric apparatus, the length of which in the fork length for *A. o. desotoi* (Vladykov 1955) was 6.8% (range 4.1–10.2%, S.D. 2.0, r = 0.062) for 12 specimens examined, and for 23 *A. o. oxyrhynchus* its mean was 8.7% with a range of 6.6–11.2% (S.D. 1.0), shows no statistically significant (p = 0.05) difference between the subspecies.

The length of the spleen as a percentage of the length of the pyloric apparatus according to Vladykov (1955) is 238.0–274.0% in *A. o. desotoi* and 30.0–98.0% in *A. o. oxyrhynchus*. In 12 *A. o. desotoi* examined the spleen length versus the pyloric apparatus length averaged 195.1% and ranged from 85.1–296.0% (S.D. 64.7, r = 0.232). In 21 *A. o. oxyrhynchus* the same morphometric ratio was 60.8% (range 29.4–110.0%, S.D. 23.7) with minimal overlap of values (Fig. 4).

There are a number of measurements in which *A. o. desotoi* and *A. o. oxyrhynchus* differ, but the differences are not so great as to be mutually exclusive. In attempting subspecific identification using morphometric data, it is necessary to use spleen length measurement ratios to verify subspecific status because of the small degree of overlap that occurs and its lesser correlation with fork length. For all three external morphometric characters that have been suggested to separate the subspecies, e.g., head length versus fork length, pectoral fin length versus fork length and dorsal scute shape, *A. o. desotoi* broadly overlaps the reference measurements obtained for *A. o. oxyrhynchus*. Although these characteristics are statistically significant (P = 0.05), no unequivocal external characters are known that will separate *A. o. desotoi* from *A. o. oxyrhynchus* 100 percent of the time.

In summary, *A. o. desotoi* and *A. o. oxyrhynchus* populations are allopatric and are sufficiently discrete and recognizable by spleen morphological characters to be considered discrete stocks in sturgeon management. Mayr (1970) defined a subspecies as: an aggregate of phenotypically similar populations of a species inhabiting a geographic

Table 5. Interior body proportions of *Acipenser oxyrhynchus oxyrhynchus* and *Acipenser oxyrhynchus desotoi*.

Body proportions	Subspecies Locality	*A. o. desotoi* Gulf of Mexico Mean (N)	*A. o. desotoi*[1] Gulf of Mexico Mean (N)	*A. o. oxyrhynchus*[1] Atlantic Ocean Mean (N)	*A. o. oxyrhynchus* Atlantic Ocean Mean (N)
Spleen length in % fork length		12.3 (12)	17.5 (2)	6.0 (13)	5.7 (22)
Spleen length in % intestine loop		76.8 (12)	101.0 (2)	31.0 (13)	39.8 (23)
Pyloric in % fork length		6.8 (12)	6.8 (2)	8.0 (13)	8.7 (23)
Spleen in % pyloric		195.1 (12)	256.0 (2)	69.0 (12)	60.8 (21)

[1] Vladykov (1955)

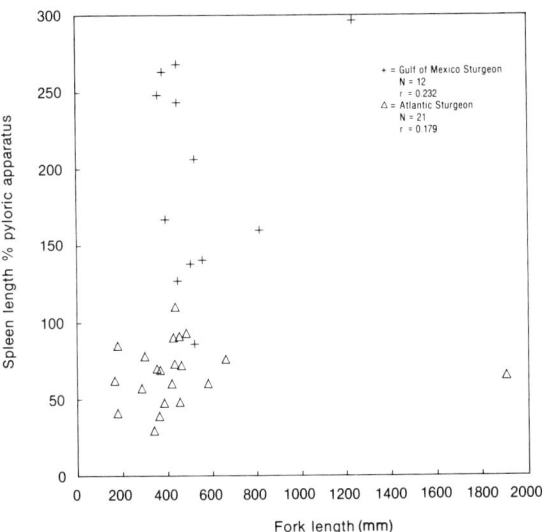

Fig. 4. Length of the spleen as a percentage of the length of the pyloric apparatus in *Acipenser oxyrhynchus desotoi* and *Acipenser oxyrhynchus oxyrhynchus*. Regression equations, line for *A. o. desotoi* (Y = 386.716X + 0.844); line for *A. o. oxyrhynchus* (Y = 391.935X + 0.972).

subdivision of the range of the species and differing taxonomically from other populations of the species. Mayr further defined taxonomic differences as differences in diagnostic morphological characters. Based on this definition and the documentation that the two taxa are differentiated in spacial distribution with only a small degree of overlap in spleen morphometrics, continued recognition of *A. o. desotoi* as a valid subspecies is supported.

Acknowledgements

I thank Edouard J. Crateau for his guidance and support in the preparation of this paper. Suggestions and critical review of this paper were provided by three anonymous reviewers. Pledger A. Moon and Frank Parauka helped with field collections. Dennis Creamer provided computer analysis assistance which was greatly appreciated. Bob Doud provided graphic assistance. Carter Gilbert and George Burgess provided specimens and facilities at the Florida State Museum, Gainesville, Florida, to study *Acipenser* collections.

References cited

Bailey, R. M. & F. B. Cross. 1954. River sturgeons of the American genus *Schaphyrhynchus*: characters, distribution, and synonomy. Pap. Michigan Acad. Sci., Arts & Letters. 39: 169–208.

Davis, J. T., B. J. Fontenot, C. E. Hoenke, A. M. Williams & J. S. Hughes. 1970. Ecological factors affecting anadromous fishes of Lake Ponchartrain and its tributaries. Louisiana Wildl. & Fish. Comm. Bull. 9: 1–63.

Gilbert, C. R. 1978. Atlantic sturgeon. pp 5–8. *In:* C. R. Gilbert (ed.) Rare and Endangered Biota of Florida, Vol. 4: Fishes, University Presses of Florida, Gainesville.

Hubbs, C. L. & K. F. Lagler. 1958. Fishes of the Great Lakes region. University of Michigan Press, Ann Arbor. 213 pp.

Huff, J. A. 1975. Life history of Gulf of Mexico sturgeon, *Acipenser oxyrhynchus desotoi*, in the Suwannee River, Florida. Florida Marine Resources Publ. No. 16. 32 pp.

Mayr, E. 1970. Populations, species and evolution. Harvard University Press, Cambridge. 797 pp.

Rivas, L. R. 1954. The origin, relationships, and geographical distribution of the Marine fishes of the Gulf of Mexico. pp. 503–505. *In:* Galtsoff, P. S. (ed.) Gulf of Mexico, Its Origin, Waters, and Marine Life, U.S. Fish and Wildlife Service, Fish. Bull. 55.

Vladykov, V. D. 1955. A comparison of Atlantic sea sturgeon with a new subspecies from the Gulf of Mexico (*Acipenser oxyrhynchus desotoi*). J. Fish. Res. Board Can. 12: 754–761.

Vladykov, V. D. & J. R. Greeley. 1963. Order Acipenseroidei. pp. 26–58. *In:* Olsen, Y. H. (ed.) Fishes of the Western North Atlantic, Sears Found. Mar. Res., Yale University, New Haven.

Wooley, C. M. & E. J. Crateau. 1982. Observations of Gulf of Mexico sturgeon (*Acipenser oxyrhynchus desotoi*) in the Apalachicola River, Florida. Florida Scientist 45: 244–248.

Wooley, C. M. & E. J. Crateau. 1985. Movement, microhabitat, exploitation and management of Gulf of Mexico sturgeon, Apalachicola River, Florida. N. Amer. J. Fish. Manage. (in press).

Received 15.5.1984   Accepted 28.3.1985

Preliminary description of the genetic structure of white sturgeon, *Acipenser transmontanus*, in the Pacific Northwest

Devin M. Bartley[1,2], Graham A.E. Gall[1] & Boyd Bentley[1]
[1] *Department of Animal Science, University of California, Davis, CA 95616, U.S.A.*
[2] *Department of Biology, San Diego State University, San Diego, CA 92110, U.S.A.*

Synopsis

This is a preliminary report on the use of starch gel electrophoresis to assess the genetic structure of four populations of white sturgeon. At least one variant allele was found at seven of nineteen loci studied. The preliminary data identify sufficient allelic variation among the populations to justify extended use of allozyme markers to distinquish populations of white sturgeon. Average heterozygosity ranged from 4.9 to 6.9% in three populations with ocean access, while a landlocked population from the Kootenai River, had a value of only 1.4%.

Introduction

The white sturgeon, *Acipenser transmontanus*, is experiencing a resurgence in popularity as a fishery resource. Commercial and academic aquaculture ventures are underway in California, Oregon, and British Columbia (Doroshov et al. 1983, Buddington & Doroshov 1984); the commercial harvest of sturgeon on the Columbia River has increased from those fish caught incidentally during salmon gill-net fishing to a specific sturgeon set-line and gill net fishery (King & Kreitman, unpublished report); and population studies aiding formation of management policies have been undertaken throughout the Pacific Northwest (Kohlhorst et al. 1980, Partridge, unpublished report, Cochnauer et al. 1985, Galbreath 1983).

As the popularity of sturgeon increases and management decisions affecting these stocks are made, knowledge of the genetic structure of sturgeon will be essential to preserve the genetic integrity and variability of white sturgeon stocks and populations. We wish to guard against the early management practices of salmonid fisheries which resulted in indiscriminate transplanting of fish (Nicola 1976) with no concern for genetic considerations. Transplantings of rainbow trout, *Salmo gairdneri*, and subsequent introgressive hybridizations with other *Salmo* species have resulted in the loss of the genetic integrity of local stocks and are presumed to be a factor in the decline of native trout populations (Busack & Gall 1981).

The purpose of this research is to develop electrophoretic techniques suitable for sturgeon allozyme analysis and to apply this technology to describe the genetic structure of four white sturgeon populations. This technique has had wide application in fish systematics, population studies and management (Allendorf & Utter 1979, Busack & Gall 1981).

F.P. Binkowski & S.I. Doroshov (ed.), North American sturgeons. ISBN 90-6193-539-3
© 1985, Dr W. Junk Publishers, Dordrecht. Printed in the Netherlands. Developments in EBF 6.

Methods and materials

Samples of white sturgeon were collected from four river systems in the Pacific Northwest: San Francisco Bay in California, the Kootenai River in Idaho, the lower Columbia River in Oregon, and the lower Fraser River in British Columbia. A one gram sample of white muscle tissue was dissected from the 'shoulder' area near the third to fourth most anterior dorsal scute. All muscle samples except those from the Kootenei River were immediately frozen on dry ice and transported to the laboratory.

Tissue preparation and standard horizontal starch-gel electrophoretic techniques are described in Busack et al. (1979). The enzyme systems and buffers used are listed in Table 1. The system of nomenclature used for loci and alleles followed the suggestions of Allendorf & Utter (1979): each locus was named using an italicized abbreviation of the enzyme name; duplicate loci were assigned hyphenated numbers with the least anodic locus designated 1, the next 2, etc. The most common allele observed at each locus was designated 100 and all other alleles were assigned numbers representative of their migration rate relative to the most common allele. Cathodically migrating alleles were designated by a negative (−) sign. For example, a variant allele at the least anodal *Pgm* locus that migrated half as far as the common allele would be designated as *Pgm*-1 (50). A variant allele at this locus that migrated twice as far as the 100 allele would be *Pgm*-1 (200).

Genetic variation was assessed by calculating average heterozygosity (H) and number of polymorphic loci. Average heterozygosity (H) was calculated from observed gene frequencies of the alleles at each locus by the expression

$$H = \sum_{j=1}^{n} \frac{(1 - \sum_{i=1}^{n} p_i^2)}{N},$$

where p_i = the frequency of the ith allele at the jth locus, n = the number of alleles at the jth locus and N = the number of loci examined.

Results

A summary of the sample sizes and the observed genetic variability is presented in Table 2. Seven of the nineteen loci surveyed were found to be polymorphic in at least one population. All seven were polymorphic in the San Francisco Bay fish while only two were polymorphic in the Kootenai River fish. The Fraser and Columbia Rivers each exhibited four polymorphic loci. Average heterozygosity ranged from a low of 0.014 for Kootenai River fish to a high of 0.069 for the Columbia River.

Table 1. Enzyme systems, abbreviations, number of loci scored and buffer systems for the analysis of muscle of white sturgeon.

Enzyme	Abbreviations	Number of loci	Buffer [1]
Adenylate kinase	Ak	2	AC
Aspartate amino transferase	Aat	1	AC
Creatine phosphokinase	Ck	2	RB.5
Glucosephosphate isomerase	Gpi	3	RB.0
Glyceraldehyde 3 phosphate dehydrogenase	Gap	2	AC
Lactate dehydrogenase	Ldh	2	RB.0
Malate dehydrogenase	Mdh	2	AC
Malic enzyme	Me	1	RB.5
Phosphoglucomutase	Pgm	2	AC
Peptidase			
leucylglycylglycine	Lgg	1	RB.0
glyclleucine	Gl	1	RB.0

[1] Buffer systems are those described by Busack et al. (1979).

The description of allozyme variability observed is limited to the expression of these isozymes in muscle, even though many isozymes are expressed in additional tissues. Diagrams of the banding patterns for the seven polymorphic loci are presented in Figures 1 and 2. Homomeric bands are designated by a solid rectangle and heteromeric bands are designated by a solid oval. Harris & Hopkinson (1976) provided an excellent discussion of enzymic banding patterns.

Aat (Fig. 1A) – Aspartate aminotransferase is a dimeric enzyme coded for by one cathodic locus in sturgeon muscle. Two anodic loci are present but are unscorable when frozen muscle is used. Variation at Aat-1 consisted of a slow migrating (−70) variant.

Ak (Fig. 1B) – Adenylate kinase is a monomeric enzyme coded for by 2 loci in sturgeon muscle. A fast migrating (140) variant was found at the Ak-2 locus.

Lgg (Fig. iC) – The peptidase, leucyl-lglycylglycine, has been reported to be a dimer, however heterodimers are not formed between the allozymes expressed in sturgeon muscle. For the present, we are scoring Lgg as a single locus with one variant (60) allele.

Mdh (Fig. 1D) – Malate dehydrogenase is a dimeric enzyme coded for by at least two loci in sturgeon. The Mdh-1 (138) variant has the same mobility as the heterodimer formed between Mdh-1 (100) and Mdh-2 (100). The presence of the Mdh-1 (138) allele can be detected by the heterodimers formed by Mdh-1 (138) and the common alleles at Mdh-1 and Mdh-2. Differences in relative staining intensities of Mdh allozymes suggested that the Mdh loci are duplicated, but this hypothesis remains untested. For the present we are

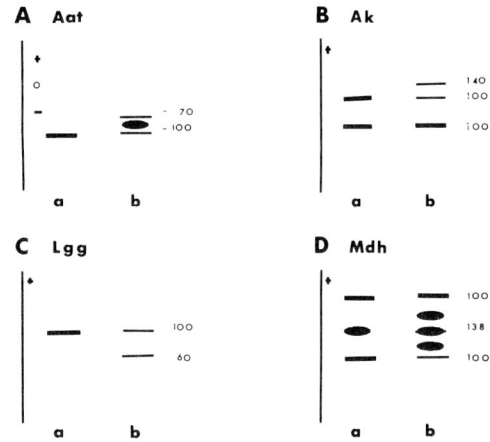

Fig. 1. Banding patterns observed for four enzyme systems from white sturgeon muscle. Homomeric bands are represented by solid bars with the allelic designation given at the right; heteromeric bands are represented by solid elipses. A – Aspartate amino transferase: (a) common (−100) homozygote; (b) −70/−100 heterozygote. B – Adneylate kinase: (a) common homozygote at both locus 1 and locus 2: (b) 140/100 heterozygote at locus 2, common homozygote at locus 1. C – Leucyl glycylglycine: (a) common homozygote; (b) 60/100 heterozygote. D – Malate dehydrogenase: (a) common homozygote at both locus 1 and 2; (b) 138/100 heterozygote at locus 1, common homozygote at locus 2. Note overlap of 100/100 heterodimer and 138 homodimeric bands.

scoring Mdh as not duplicated.

Gpi (Fig. 2A) – Glucosephosphate isomerase is a dimeric enzyme coded for by three loci in sturgeon muscle. Variation was found at Gpi-3 and consisted of both fast (120) and slow migrating (78) variants. The Gpi-3 (78) variant has the same mobility as the heterodimer formed between Gpi-3 (100) and Gpi-2 (100). The presence of the Gpi-3 (78) allele can be detected by the heterodimer it forms with Gpi-3 (100).

Pgm (Fig. 2B) – Phosphoglucomutase is a mono-

Table 2. Number of fish sampled and the genetic variability observed for four populations of white sturgeon. A total of 19 loci was examined.

Area	Number of fish	Polymorphic loci	Heterozygosity
San Francisco Bay	64	7	.049
Kootenai River	28	2	.014
Colombia River	28	4	.069
Fraser River	10	4	.053

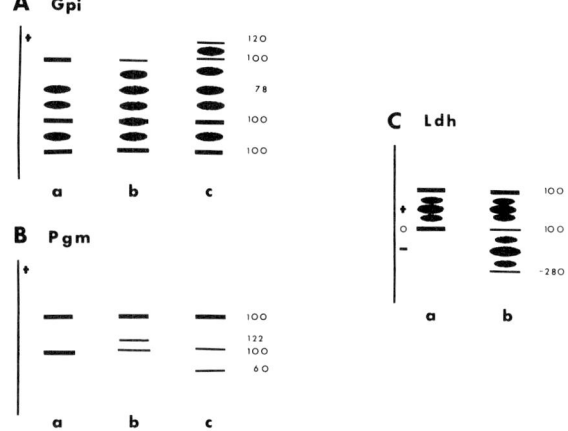

Fig. 2. Banding patterns observed for three enzyme systems from white sturgeon muscle. Homeric bands are represented by solid bars with the allelic designation given at the right; heteromeric bands are represented by solid elipses. A – Glucose phosphate isomerase: (a) common homozygotes at locus 1, 2, and 3, interlocus heterodimers; (b) 78/100 heterozygote at locus 3, locus 1 and 2 are homozygous; (c) 120/100 heterozygote. B – Phosphoglucomutase: (a) common homozygote at locus 1 and 2; (b) 122/100 heterozygote at locus 1, common homozygote at locus 2; (c) 60/100 heterozygote at locus 1, common homozygote at locus 2. C – Lactate dehydrogenase: (a) common homozygote at locus 1 and 2, interlocus heterotetramers formed; (b) −280/100 heterozygote at locus 1, common homozygote at locus 2, no heterotetramers formed between the products of the −280 allele and the 100 allele at locus 2.

meric enzyme coded for by 2 loci in sturgeon muscle. Variation consisted of the (122) and (60) alleles at Pgm-1.

Ldh (Fig. 2C) – Lactate dehydrogenase is a tetrameric enzyme coded for by 2 loci in adult sturgeon. Ldh-1 (100) migrates slightly cathodically while Ldh-2 is anodic; heterotetramers are formed between Ldh-1 and Ldh-2. A rapidly migrating Ldh-1 (−280) variant was found in all stocks.

The frequencies of the variant alleles in each population are presented in Table 3. The San Francisco Bay fish had four variant alleles, Aat-1 (−70), Gpi-3 (120) Ak-2 (140), and Lgg (60), not found in fish from other populations. This may be an artifact of the small sample size of 3 of the populations. Conversely, the San Francisco fish did not possess the Pgm-1 (60) allele observed in the other 3 populations.

Discussion

The average level of heterozygosity observed for fish from the four sturgeon populations was comparable to levels reported for salmonid species except rainbow trout, but higher than those previously reported for members of Acipenseriformes. Five species of Pacific salmon have values ranging from 1.5% to 4.5% (Allendorf & Utter 1979). Paddlefish, *Polyodon spathula*, from the Mississippi River were monomorphic at 33 of 35 loci and the average heterozygosity was 0.013 (Carlson et al. 1982). No electrophretic differences were found for two species of *Scaphirhynchus*, the pallid, *S. albus*, and the shovelnose, *S. platorynchus*, sturgeon, at 37 loci examined (Phelps & Allendorf 1983).

Table 3. Frequencies of the variant alleles at seven polymorphic loci observed in four populations of white sturgeon. The frequency of the common allele, designated 100, has not been included in the table.

Locus	Allele	S.F. Bay	Kootenai	Columbia	Fraser
Ak-2	140	.08	.00	.00	.00
Aat-1	−70	.08	.00	.00	.00
Ldh-1	−280	.03	.11	.08	.11
Mdh-1	138	.02	.00	.14	.10
Gpi-3	120	.02	.00	.00	.00
	78	.22	.00	.32	.10
Pgm-1	122	.05	.02	.09	.25
	60	.00	.02	.03	.05
Lgg	60	.08	.00	.00	.00

lendorf & Utter 1979). Therefore, the low value for the Kootenai River fish may be a reflection of a bottleneck effect resulting from the construction of Libbey Dam isolating this population or from limited reproductive population sizes since construction.

The allele frequency data suggest genetic differences among the four populations. The presence of the *Aat*-1 (−70), *Ak*-2 (140) and *Lgg* (60) alleles exclusively in the San Francisco Bay fish and the absence of the *Pgm*-1 (60) allele from these fish indicates a distinct allelic distribution among these populations. In populations of the American lobster, *Homarus americanus*, and the Malaysian prawn, *Macrobrachium rosenbergii*, discrete subpopulations were evident from single allele differences among the populations even though average heterozygosity was low (Hedgecock et al. 1979, Tracy et al. 1975).

Although white sturgeon appear to be morphologically conservative, the fish from the four populations studied also show slight differences in girth, growth rate, spawning time, color, and snout length (Semakula & Larkin 1968, Kohlhorst et al. 1980, Jim Lukens, personal communication, and personal observation). As yet we do not know to what extent these differences are influenced by genetic or environmental effects. We believe electrophoretic analysis, used along with other management assessment techniques, will provide valuable information on which to base fishery management decisions concerning the genetic resource of sturgeon populations.

References cited

Allendorf, F.W. & F.M. Utter. 1979. Population genetics. pp. 407–454 *In*: J. Hoar, D.J. Randall & J.R. Brett (ed.) Fish Physiology, Vol 8, Academic Press, New York.

Buddington, R.K. & S.I. Doroshov. 1984. Feeding trials with hatchery produced white sturgeon juveniles (*Acipenser transmontanus*). Aquaculture 36: 237–243.

Busack, C.A. & G.A.E. Gall. 1981. Introgressive hybridzation in populations of Paiute cutthroat trout. Can. J. Fish. Aquat. Sci. 38: 939–951.

Busack, C.A., R. Halliburton & G.A.E. Gall. 1979. Electrophoretic variation and differentiation in four strains of domesticated rainbow trout. Can. J. Genet. Cytol. 21: 81–94.

Carlson, D.M., M.K. Kettler, S.E. Fisher & G.S. Whitt. 1982. Low genetic variability in paddlefish populations. Copeia 1982: 721–725.

Cochnauer, T.G., J.R. Lukens & F.E. Partridge. 1985. Status of white sturgeon *Acipenser transmontanus*, in Idaho. Dev. Env. Biol. Fish. 6: 00–00.

Doroshov, S.I., W.H. Clark, Jr., P.B. Lutes, R.L. Swallow, K.E. Beer, A.B. McGuire & M.D. Cochran. 1983. Artificial propagation of the white sturgeon, *Acipenser transmontanus*. Aquaculture 32: 93–104.

Galbreath, J.L. 1983. Status, management and life history of Columbia River white sturgeon, *Acipenser transmontanus*. 113th Meeting American Fisheries Society, Milwangee (Abstract).

Harris, H. & D.A. Hopkinson. 1976. Handbook of enzyme electrophoresis in human genetics. North-Holland Publishing Co., Amsterdam. 281 pp.

Hedgecock, D., D.J. Stelmach, K. Nelson, M.E. Lindfelser & S.R. Melecha. 1979. Genetic divergence and biogeography of natural populations of *Macrobrachium rosenbergii*. Proc. World Mariculture Soc. 10: 873–879.

Kohlhorst, D.W., L.W. Miller & J.J. Orsi. 1980. Age and growth of white sturgeon collected in the Sacramento-San Joaquin estuary, California: 1965–1970 and 1973–1976. Calif. Fish & Game 66: 83–95.

Nicola, S.J. 1976. Fishing in Western national parks – a tradition in jeopardy? Fisheries 1: 18–21.

Phelps, S.R. & F.W. Allendorf. 1983. Genetic identity of pallid and shovelnose sturgeon. Copeia 1983: 696–700.

Semakula, S.N. & P.A. Larkin. 1968. Age, growth and yield of the white sturgeon of the Fraser River, British Columbia. J. Fish. Res. Board Can. 25: 2589–2602.

Tracy, M.L., K. Nelson, D. Hedgecock, R.S. Shlesser & M. Pressick. 1975. Biochemical genetics of lobsters: genetic variation and structure of American lobster, *Homarus americanus*, populations. J. Fish. Res. Board Can. 32: 2091–2101.

Received 15.5.1984   Accepted 10.3.1985

Habitat use and behavior of pre-spawning and spawning shortnose sturgeon, *Acipenser brevirostrum*, in the Connecticut River

Jack Buckley[1] & Boyd Kynard
Massachusetts Cooperative Fishery Research Unit, Department of Forestry and Wildlife Management, University of Massachusetts, 204 Holdsworth Hall, Amherst, MA 01003, U.S.A.
[1] *Present address: Department of Consumer and Regulatory Affairs, Water Quality Branch/Fisheries, 5010 Overlook Ave., S.W., Washington, D.C. 20032, U.S.A.*

Keywords: Reproduction, Telemetry, Ecology, Movements, Spawning site, Chondrostean fish

Synopsis

Movements and ecology of pre-spawning and spawning shortnose sturgeon, *Acipenser brevirostrum*, were studied through 1979–1982 in a 2 km reach of the Connecticut River. Radio telemetry was used to monitor the movements of 18 sturgeon. An additional 165 sturgeon, captured by gillnets, provided information on spawning site selection, sex ratio, and reproductive condition. For 3 years the mean water velocities during the spawning period ranged from 0.36 to 1.2 m sec^{-1} in the spawning area. Substrate was cobble and rubble. Sturgeon spawned over a short time period (3–5 days), during decreasing river discharge of 679 to 301 m^3 sec^{-1} ($\bar{x}$ = 561 m^3 sec^{-1}) and rising water temperature between 11.5 to 14.0° C. High river discharge over a prolonged period during the normal spawning season may preclude reproduction.

Introduction

Information on the location and timing of sturgeon spawning in North America lacks specific details and is based on captures of reproductively mature fish or extrapolation from collection of larvae. The location and timing of spawning has been determined using larval collections for white sturgeon, *Acipenser transmontanus* (Kohlhorst 1976), shovelnose sturgeon, *Scaphirhynchus platorynchus* (Helms 1974, Moos 1978), and with the capture of mature adults for lake sturgeon, *A. fulvescens* (Harkness & Dymond 1961) and Atlantic sturgeon, *A. oxyrhynchus* (Huff 1975, Dovel 1977). Taubert (1980) used the gillnet capture of mature adults and 13 larvae to back calculate the general upriver area, physical conditions of the river, and the timing of spawning for a landlocked population of shortnose sturgeon, *A. brevirostrum*, upstream of Holyoke Dam in the Connecticut River. Dadswell (1979) also used the capture of adults and larvae in the Saint John River, New Brunswick to estimate the timing and general location of spawning. Although this information is useful, the lack of specific details about the spawning sites are due to the limitations of the methodologies used in these studies.

The present study was conducted in a spawning area that was located during an investigation of the annual movement of shortnose sturgeon (Buckley 1982). Shortnose sturgeon spawned below Holyoke Dam each year in May but most fish, that would spawn in the spring migrated to the spawning area in the early fall and overwintered in the area. Thus, pre-spawning fish were in the spawning

---

* Contribution 92 of the Massachusetts Cooperative Fishery Research Unit which is supported by the U.S. Fish and Wildlife Service, Massachusetts Division of Fisheries and Wildlife, Massachusetts Division of Marine Fisheries, and the University of Massachusetts.

F.P. Binkowski & S.I. Doroshov (ed.), North American sturgeons. ISBN 90-6193-539-3
© 1985, Dr W. Junk Publishers, Dordrecht. Printed in the Netherlands. Developments in EBF 6.

area for about 8 months prior to spawning. We studied the movements and habitat use of fish during the pre-spawning and spawning periods. Radio telemetry and gillnet captures were used to determine movement and gillnet captures to determine the reproductive condition and sex of fish. Concurrent with telemetry and gillnetting, information was collected on the habitat in the spawning area, i.e., substrate, water velocity, and water depth.

Study area

The study area was the 2 km section of the Connecticut River immediately below Holyoke Dam (Fig. 1). The dam, located at river km 140, is a 9 m high hydroelectric dam that diverts the river into a hydroelectric power station and extensive canal system. The dam was constructed in 1848 at the site of a natural falls (Green 1939). A fish elevator passes several species of anadromous fish upriver each spring and fall (Moffitt et al. 1982) but shortnose sturgeon do not typically use the fish elevator (35 passed since 1955, Massachusetts Division of Fisheries and Wildlife data). Included in the study area was the 800 m long discharge tailrace of the power station (Fig. 1). The river parallel to the tailrace is dewatered during periods of low discharge ($<357 \text{ m}^3\text{sec}^{-1}$). During this period the entire river flow is usually diverted through the power station which discharges into the tailrace, or into a canal system that discharges downriver of the study area. Thus, during low discharge periods flow from the tailrace is the only flow through the study area.

Materials and methods

Physical characteristics of study area

River discharge and water temperature at Holyoke Dam were determined using information supplied by Holyoke Water Power Company, Holyoke, Massachusetts. For the survey of substrate, water depth, and water velocity, we established perpendicular transects across the river at six canal exits on the west bank (Fig. 1). Five stations in 1982 and three in 1983 were located equidistant along each

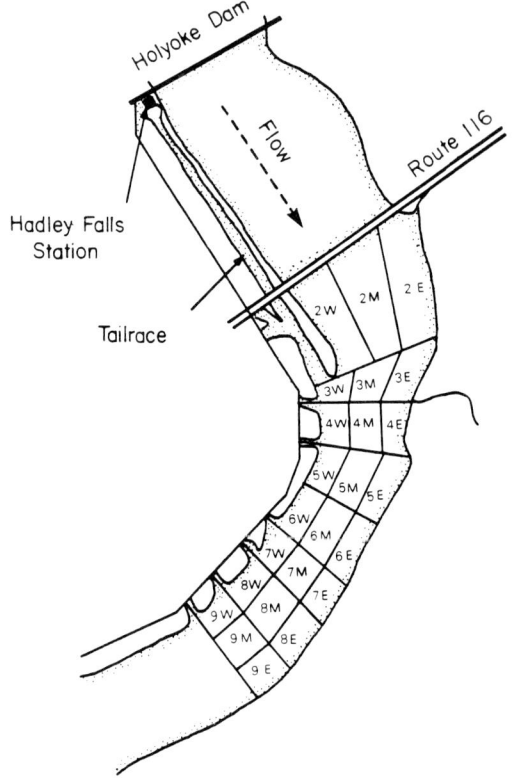

Fig. 1. The 2 km study area of the Connecticut River immediately below Holyoke Dam, Holyoke, Massachusetts, showing transects and grid system used for habitat surveys and location of radio tagged fish.

transect using a rangefinder accurate to 1 m at 100 m. At each station, velocity and depth were measured and a substrate sample was collected. Velocity was measured at 0.6 of the water depth using a Marsh-McBirney current meter. Substrate was collected with an Ekman dredge and classified in the field according to the dominant material as silt, sand, gravel (≤ 25 mm), cobble (27–75 mm), or rubble (≥ 75 mm). Classification of the substrate was confirmed by visual inspection during the summer low-discharge period.

Gillnet captures

Fish were captured each spring from 1980 through 1983 and in the fall of 1981 and 1982. We used 100 m bottom set nets with stretch mesh of 10.0, 12.5, 17.5 and 20.0 cm; fishing effort was distributed evenly throughout the study area. Fish were measured for fork length (FL) to the nearest mm and weighed with a spring scale, which was accurate to ±50 g. A numbered Atkins tag was attached with stainless steel wire through a hole drilled in a dorsal scute. The sex and reproductive condition of fish was determined by visual inspection.

Telemetry monitoring

Radio telemetry was used to monitor the habitat use of fish from October 1979 through May 1980 and October 1982 through May 1983. Fish were either tagged downstream of Holyoke Dam during summer in feeding areas or in the spawning area during the fall. Transmitter attachment was similar to that used to attach Atkins tags except that two wires were used to hold the transmitters on the fish. The cylindrical transmitters were semibuoyant, weighed between 10 and 15 g in water, and measured 6 × 1.5 cm. No tag exceeded 1.25% of a fish's body weight. Individual fish were identified by frequency and pulse rate. Fish were located from a boat.

From October 1979 through April 1980 fish were located biweekly prior to spawning and located daily during the spawning period. From October 1982 through 14 April 1983, fish were located weekly; from 15 April through 15 May, fish were located each hour during daylight for 2 days and each hour for 24 hours on the third day.

To facilitate plotting the telemetry observations a grid system was developed based on the transects that were used for surveying habitat. Each grid area was defined up and downriver by a transect that crossed the river, and the position of the grid along the transect was designated as east (east one-third), west (west one-third) or mid-river (middle one-third) (Fig. 1). Two additional areas were designated as part of the grid system; the power station tailrace, and the area between the tailrace peninsula and the east bank of the river.

The locations of fish determined by telemetry and gillnet captures were classified into pre-spawning or spawning periods. Our indirect determination of the period when spawning most likely occurred was based on movements and activity of fish, the temperature and discharge of the river, the reproductive condition of captured fish, and the departure of radio tagged fish from the study area. Thus, our reference to the prespawning and spawning periods is based on a combination of factors; we collected no fertilized eggs or emergent larvae.

Results

Physical characteristic of spawning area

Seventy-five percent of the substrate samples were classified as imbedded gravel, cobble, and rubble. There were deposits of sand and silt in two areas, one area associated with the downstream growth of the tailrace peninsula, and the second caused by an eddy in areas 5 W and 6 W (Fig. 1).

The most distinct feature of bottom topography was a depression located downriver of the entrance to the tailrace (Fig. 2A). This is the deepest water available for a distance of 3 km below the dam.

Water velocities were measured at river discharges of 301, 670, and 1020 m^{3}sec^{-1} (Fig. 2B, C, D). These discharges represent a range of typical river discharges that are present before and during spawning.

Timing of spawning

In 1980, 1981, and 1982 spawning occurred with water temperature increasing from 11.5 to 14° C and discharge declining from 679 to 301 m^{3}sec^{-1} ($\bar{x}$ = 561 m^3 sec^{-1}). In 1980 spawning occurred from 1 through 6 May, in 1981 from 30 April through 5 May, and in 1982 from 5 through 10 May.

Gillnet captures

During the three years of sampling 165 shortnose sturgeon were captured. These fish weighted 0.1 to 9.2 kg and were 53 to 97 cm FL. We determined the sex of 49 fish, 15 females and 34 males.

During the winter and spring (pre-spawning period) 69 fish were captured. River discharge during this period was <340 m^{3}sec^{-1}. Most captures were in the slowest available velocity (Fig. 2B, 3A).

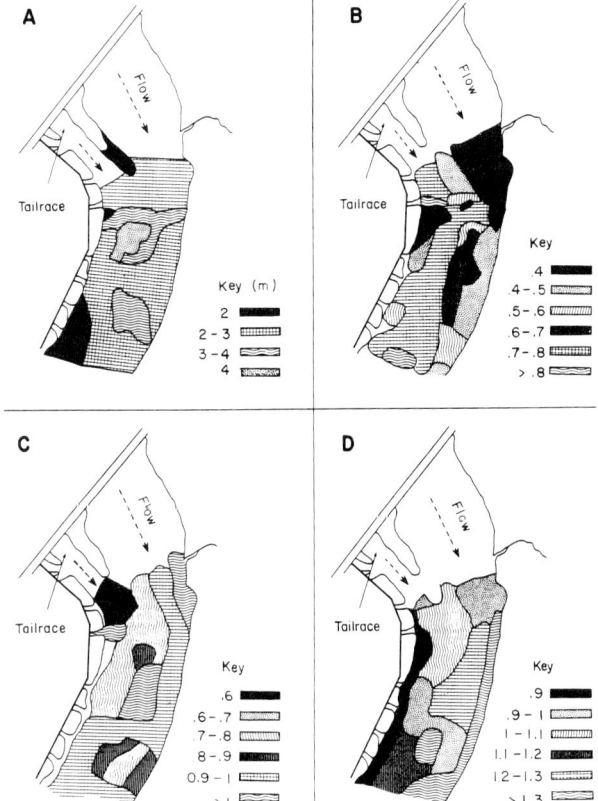

Fig. 2. Water depths and water velocities in the study area: A – depth (m) profile at discharge of $301\,m^3\,sec^{-1}$; B – water velocities $(m\,sec^{-1})$ at discharge of $301\,m^3\,sec^{-1}$; C – water velocities $(m\,sec^{-1})$ at $670\,m^3\,sec^{-1}$; D – water velocities $(m\,sec^{-1})$ at $1030\,m^3\,sec^{-1}$.

During the spawning period 96 fish were captured. Forty were easily sexed and the sex ratio was 33 males and 7 females (17.5% females). The majority were captured in water velocities from 0.7 to $1.0\,m\,sec^{-1}$. When these fish were captured the available water velocities were 0.3 to $1.2\,m\,sec^{-1}$. Most fish were captured along the western shore or in the deep area downriver from the mouth of the tailrace (Fig. 3B). The reproductive condition of the 19 fish captured at the downriver edge of the study area (9W) indicated they had already spawned and were leaving the area (Fig. 3B). These fish were captured in 1982 on two consecutive days and two were spent females. Only four days before their departure, 7 fish were captured with running eggs and sperm. Thus, the spawning period appeared to last only 3 to 5 days in 1982.

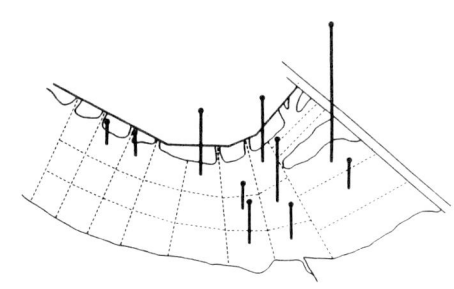

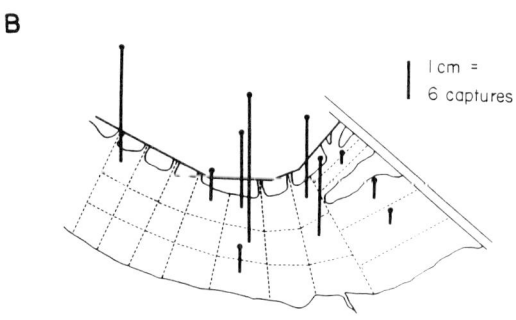

Fig. 3. Gillnet captures in the study area: A – captures during the pre-spawning periods of 1981 and 1982; B – captures during the spawning periods 1980, 1981 and 1982.

During the pre-spawning period most captures of males were of isolated individuals. In contrast, males captured during the spawning period were in groups of 2 to 15 with 1 or 2 females.

Telemetry monitoring

In the pre-spawning period of 1980, 10 fish were located 161 times. One-third of the locations were in the tailrace (Fig. 4A). Fish were most frequently located in the upriver transects where velocity was low because there was no spillage over the dam. This was similar to the results from gillnet captures (Fig. 3A). Fish moved into areas of increased velocity during the spawning period in 1980 (Fig. 4B), which was also similar to the results from gillnet captures (Fig. 3B).

In the spring of 1983, 8 pre-spawned fish were located 206 times. River discharge was very high i.e., 914 to $1800\,m^3\,sec^{-1}$ (mean of $1400\,m^3\,sec^{-1}$). Twenty-one percent of the locations were in the tailrace (Fig. 4C). Other locations of fish were

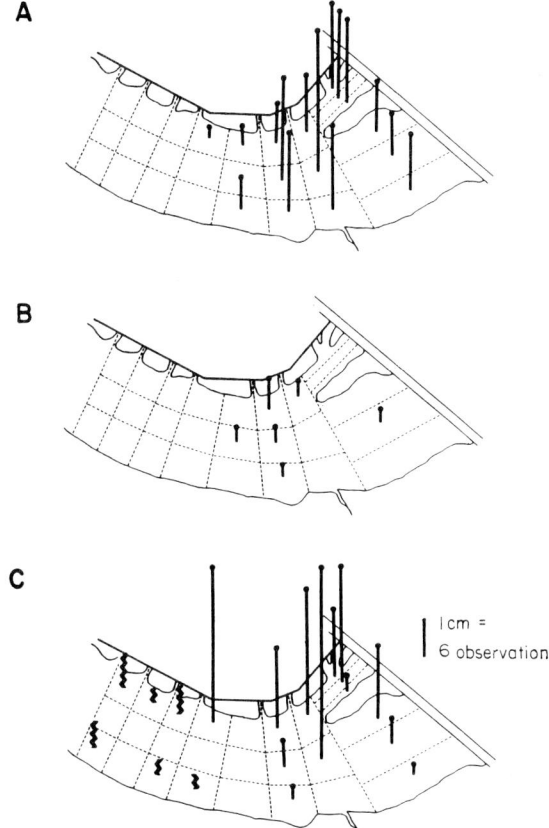

Fig. 4. Telemetry observations: A – pre-spawning period in 1980; B – spawning period in 1980; C – spring 1983. The locations indicated by the vertical zigzag bars were fish that were leaving the study area.

concentrated in areas of reduced velocity below the tailrace penninsula or along the western shore of the river. Fish located in the lower sections of the study area were fish that were leaving the study area (Fig. 4C). By 6 May all fish had left the study area and the river discharge was still high (1600 m^3 sec^{-1}). When fish left the area they did not rapidly move downriver as post-spawned fish did in 1980 (Buckley 1982). These fish only moved downriver about 2 km into an area of reduced velocity and remained for 2 weeks. They were solitary and did not behave as the spawning sturgeon did in 1980. After 2 weeks they moved further downriver into summer feeding areas. The shortnose sturgeon never appeared to spawn in 1983.

Discussion

Telemetry monitoring and gillnet captures showed a clear pattern of habitat use in the winter/spring pre-spawning period that appeared related to changes in river discharge. Pre-spawning shortnose sturgeon preferred areas of reduced velocities, between 0.3 and 0.7 m sec^{-1}. When the tailrace discharge determined the velocity patterns (low discharge conditions), fish used different areas than when spillage over the dam determined the velocity pattern (high discharge conditions). The movement of fish seemed an attempt to remain in areas of lower velocity.

Several studies have found that shortnose sturgeon spawn in areas where the general substrate is gravel, rubble, and cobble (Dadswell 1979, Taubert 1980, Squires 1982). The widespread availability of suitable spawning substrate throughout our study area suggest that water velocity and depth may be more critical than substrate factors in determining the specific spawning location. The release of eggs in the proper velocity is critical to successful egg deposition and survival. If eggs are released when the velocity is too high, the eggs may not adhere to the substrate before losing adhesiveness. Spawning in insufficient velocity may cause clumping of eggs which would increase mortality from respiratory stress, fungus growth, and, possibly increased egg predation. Spawning in sufficient water velocity would also be important to survival of larval. Buckley & Kynard (1981) described newly hatched shortnose sturgeon embryos as active and engaging in a series of vertical swimming bouts. Similar behavior has been reported for several other species of sturgeon (Soin 1971, Balon 1975). Presumably this is an adaption for downriver transport, and, if hatching does not occur in an area of sufficient velocity, the hatching fish would not be transported downstream.

The telemetry and gillnet capture data in 1980 indicated spawning was associated with a deep depression in the river bottom. Squires et al. (1981) reported that shortnose sturgeon spawned in a deep area in the Androscoggin River in Maine. By spawning in a depression, females would have the greatest range of depths and, thus, velocities that were available.

Although the association of several males with one female during spawning is based on inference from gillnet captures rather than direct observations, similar behavior has been observed in the lake sturgeon (Priegel & Wirth 1975, Harkness & Dymond 1961). It appears that a single female is accompanied by several males during spawning; this would be typical spawning behavior for a lithophilous species.

Contrary to Harkness & Dymond (1961), we found fish did not abandon the spawning area when captured and handled. This was verified with several radio tagged fish which were captured by gillnet, released and remained in the area.

Shortnose sturgeon, lake sturgeon, and several species in the Soviet Union have a compressed spawning period, perhaps no more than 3 to 5 days (Stone 1910, Harkness & Dymond 1961, Khoroshko 1972, Taubert 1980). This is strongly supported by our results. It appears that environmental conditions suitable for spawning are available for only a short period.

Water temperatures during spawning were similar to those reported for shortnose sturgeon by Dadswell (1979) and Taubert (1980). There are few reports of the water velocities at sturgeon spawning sites. Volinov & Kasyanov (1976) reported *Acipenser baeri* spawned in velocities from 0.76 to 1.53 m sec^{-1}, which are similar to the velocities that we found for shortnose sturgeon.

Water temperature and water velocity may affect the reproductive process in a stepped manner. Rising water temperatures would cause the final maturation of oocytes and the appropriate water velocity may cue the female to deposit eggs. There is some evidence in fishes that the yearly reproductive rhythms are under control of an endogenous circannual (or longer) mechanism, mediated by environmental conditions that control recrudescence and regression of gonads (de Vlaming 1972, Baggerman 1979). In the context of life history strategy the relationship between endogenous and exogenous factors is further complicated in an iteroparous species like sturgeon. In these species, if environmental conditions are not suitable, presumably natural selection would favor reabsorbtion of gonadal tissue. Two studies have reported gonadal regression in sturgeon under unfavorable environmental conditions (Kozlovsky 1968, June 1977).

Our telemetry monitoring during the period when fish should have spawned in 1983 indicated that environmental conditions were not suitable for spawning before endogenous factors precluded reproduction. The departure of fish from the spawning area before they had spawned suggests a threshold that is defined by the duration and magnitude of water velocity. Although our results are preliminary, this could represent a seldom recognized phenomenon in iteroparous lithophilic fish.

Acknowledgements

We would like to extend special thanks to Fred Binkowski and Sergei Doroshov for organizing the Sturgeon Symposium. H.E. Booke and W.E. Dodge provided constructive comments of earlier drafts on the manuscript. Our thanks go to the following field personnel, whose efforts under difficult conditions made this work possible: S. Hurley, J. O'Leary, T. Clifford, A. Richmond, S. Quinn, K. Bugley, and J. Nicholson. Financial support was provided by the National Marine Fisheries Service, the Connecticut River Watershed Council, and Northeast Utilities Service Company.

References cited

Baggerman, B. 1979. Photoperiodic and endogenus control of the annual reproductive cycle in teleost fishes. pp. 533–567. *In:* M.A. Ali (ed.) Environmental Physiology of Fishes, Plenum Press, New York.

Balon, E.K. 1975. Reproductive guilds in fishes: a proposal and definition. J. Fish. Res. Board Can. 32: 821–864

Buckley, J. & B. Kynard. 1981. Spawning and rearing of shortnose stureon from the Connecticut River, Progr. Fish-Cult. 43: 74–76.

Buckley, J.L. 1982. Seasonal movement, reproduction, and artificial spawning of shortnose sturgeon (*Acipenser brevirostrum*) from the Connecticut River. M.S. Thesis, University of Massachusetts, Amherst. 64 pp.

Dadswell, M.J. 1979. Biology and population characteristics of the shortnose sturgeon, *Acipenser brevirostrum*, Le Suer (Osteichthyes: Acipenserdae) in the Saint John River estuary, New Brunswick, Canada. Can. J. Zool. 57: 2186–2210.

de Vlaming, V.L. 1972. Environmental control of teleost reproductive cycles: a brief review. J. Fish Biol. 4: 131–140.

Dovel, W.L. 1977. Biology and management of shortnose sturgeon and Atlantic sturgeon of the Hudson River. Report for New York Department of Environmental Conservation, Albany.

Green, C. 1939. Holyoke, Massachusetts—A case history of the industrial revolution in America. Yale University Press, New Haven. 425 pp.

Harkness, W.J.K. & J.R. Dymond. 1961. The lake sturgeon. Ont. Dept. Lands Forests, Maple. 97 pp.

Helms, D. 1974. Shovelnose sturgeon in the Mississippi River, Iowa. Tech Series 74-3, Iowa Dept. of Conserv. 68 pp.

Huff, J.A. 1975. Life history of the Gulf of Mexico sturgeon, *Acipenser oxyrhynchus desotoi*, in the Suwannee River, Florida. Flor. Mar. Res. Publ. 16: 1–32.

June, F.C. 1977. Reproductive patterns in seventeen species of warmwater fishes in a Mississippi reservoir. Env. Biol. Fish. 2: 285–296.

Khoroshko, P.N. 1972. The amount of water in the Volga Basin and its effect on the reproduction of sturgeons (Acipenseridae) under conditions of normal and regulated discharge. J. Ichthyol. 12: 608–616.

Kohlhorst, D.W. 1976. Sturgeon spawning in the Sacramento River in 1973, as determined by distribution of larvae. Calif. Fish and Game 62: 32–40.

Kozlovsky, D.A. 1968. Resorption of sexual products in fishes as a stimulus to biological modification. Prob. Ichthyol. 8: 803–807.

Moffitt, C.M., B. Kynard & S.G. Rideout. 1982. Fish passage facilities and anadromous fish restortion in the Connecticut River Basin. Fisheries 7(6): 2–10.

Moos, R.E. 1978. Movement and reproduction of shovelnose sturgeon, *Scaphirynchus platorynchus*, in the Missouri River, South Dakota. Ph.D. Dissertation, University of South Dakota, Vermillion. 216 pp.

Priegel, G.R. & T.L. Wirth. 1975. The lake sturgeon: its life history, ecology and management. Wisconsin Dept. Nat. Resources Publ. Madison 240–70. 19 pp.

Soin, S.G. 1971. Adaptational feature in fish ontogeny. Israel Program for Scientific Transl., Jerusalem. 72 pp.

Squires, T.S. 1982. Evaluation of the 1982 spawning run of shortnose sturgeon (*Acipenser brevirostrum*) in the Androscoggin River, Maine. Final report to Maine Department of Marine Resources. 14 pp.

Squires, T.S., L. Flagg, M. Smith, K. Sherman & D. Ricker. 1981. American shad enhancement and status of sturgeon stocks in selected Maine waters. Annual Progress Report Maine Depart. Marine Resources, Augusta. 48 pp.

Stone, L. 1910. The spawning habits of the lake sturgeon. Trans. Amer. Fish. Soc. 29: 118–123.

Taubert, B.D. 1980. Reproduction of shortnose sturgeon, *Acipenser brevirostrum*, in the Holyoke Pool, Connecticut River, Massachusetts. Copeia 1980: 114–117.

Volinov, N.P. & V.P. Kasyanov. 1976. The ecology and reproductive efficiency of the Siberian sturgeon, *Acipenser baeri*, in the Ob as affected by hydraulic engineering works. J. Ichthyol. 18: 25–35.

Received 15.5.1984   Accepted 27.2.1985

Status, life history, and management of Columbia River white sturgeon, *Acipenser transmontanus*

James L. Galbreath
Oregon Department of Fish and Wildlife, Columbia Region Headquarters, 17330 S. E. Evelyn Street, Clackamas, OR 97015, U.S.A.

Keywords: Exploitation, Regulations, Stock, Harvest, Age, Growth, Food, Tagging, Migration, Recruitment

Synopsis

Exploitation of Columbia River sturgeon in the 1860–1890s caused severe depletion of the stocks. Stringent fishery regulations were promulgated to protect the resources, including minimum-maximum size limits of 91–183 cm TL for sport and 122–183 cm TL for commercial. Regulations, increased food supplies, and shortened salmon gill-net seasons are primary reasons for a healthy stock in the Columbia River below Bonneville Dam. The recent 10 year catch levels are the highest of the century with average annual harvest of near 40 000 fish, with a 2–3:1 ratio of sport to commercial landings. White sturgeon below Bonneville Dam grow at an annual mean rate of 6.6 cm with a range of 2.1–14.0 cm between ages 1–21 years. A current tagging program suggests that lower Columbia River sturgeon between 70 and 130 cm TL grow an average of about 10 cm per year. White sturgeon enter the sport fishery at a minimum length of 91 cm and a mean age of 9 years, the commercial fishery at a minimum length of 122 cm and a mean age of 12 years, and leave both fisheries at a maximum length of 183 cm and a mean age of 20 years. Males mature at 12 years and females 15–20 years. White sturgeon above Bonneville Dam are essentially landlocked within each successive pool. Stock size and recruitment appear satisfactory in some areas, but decrease is evident in others.

Introduction

White sturgeon, *Acipenser transmontanus*, and green sturgeon, *A. medirostris* are found in the Columbia River. White sturgeon are found throughout the Columbia River with the greatest numbers below Bonneville Dam. They have high flesh quality and are a very desirable sport and commercial fish. In contrast green sturgeon are primarily estuarine, and because of inferior flesh quality and low catchability, are of minor importance to the commercial and sport fisheries. On rare occasions green sturgeon are found as far upstream as Bonneville Dam, particularly during low flow years when saltwater intrusion is greatest.

Commercial sturgeon fisheries are managed jointly by the Washington Department of Fisheries (WDF) and the Oregon Department of Fish and Wildlife (ODFW) under authority of the Columbia River Compact established in 1918. The states manage the sport fishery separately. The primary management activities are monitoring fisheries, sampling catches and larvae, and a tagging program. There has not been a large-scale research project on Columbia River sturgeon since the late 1940s and early 1950s (Bajkov 1949, 1951). However, with the salmonid problems in the Columbia River and Pacific Ocean and the increasing importance of sturgeon as a food and game fish, more interest is now shown in sturgeon research.

Background

Sturgeon were dominant when the first salmon gill-net fisheries began on the Columbia River in 1860. At that time sturgeon were routinely killed and discarded in an attempt to eradicate them. However, by 1880 an important industry on the Columbia River had begun, with 85 m tons of salted and pickled sturgeon used locally, and the first railroad car of frozen sturgeon shipped to the East Coast and San Francisco (Craig & Hacker 1940).

Several types of gear were used to catch sturgeon including fish wheels, traps, gill nets, baited setlines and Chinese gang lines. By 1892 a peak production of 2500 tons was reached. The average weight of sturgeon was about 68 kg, but by 1895 had dropped to 25 kg. The fishery collapsed in the late 1890s due to overfishing, particularly on the large female brood stock, and landings had dropped to 33 100 kg (Fig. 1).

Sturgeon catches from the 1900s through the 1960s were small and incidental to salmon gill-net fishing and a further restriction on the fishery became necessary. As a result of extensive research by Oregon Fish Commission (OFC) biologists, the most important protective regulation was enacted in 1950, a maximum size limit of 183 cm to protect the female brood stock.

Also in 1950, a minimum size limit of 76 cm and a five fish per day bag limit for sturgeon anglers was adopted. In 1951 the bag limit was reduced to three fish per day. In 1958 the minimum length was increased to 91 cm. With the minimum and maximum size limits, reduced salmon gill-net seasons, food supplies, and pollution cleanup, the sturgeon stock below Bonneville flourished. Commercial sturgeon landings almost doubled in the early 1970s, nearly doubled again in the late 1970s, and have remained at a reduced level in the early 1980s (Table 1).

In terms of numbers of fish, the recent peak catch of 1983 was greater than the historical peak catch of 1892. However, the average weight now is 14–16 kg in the commercial fishery and about 8 kg in the sport fishery.

Fisheries below Bonneville

The majority of the commercial catch of sturgeon is incidental during salmon gill-net seasons. However, some fishermen target on sturgeon with large-mesh nets in areas of known abundance.

Landings of sturgeon during salmon gill-net fisheries in the main-stem Columbia River below Bonneville Dam during 1975–1983 varied with sturgeon abundance and salmon fishery duration. Salmon gill-net fishing averaged 80 days annually from 1960–1974. The average decreased to 48 days in the period 1975–1979. Reduced commercial sturgeon landings in the 1980s are probably a result of shortened salmon seasons, which have averaged 27 days.

Because of declining salmon stocks and commercial gill-net fishing opportunity, experimental setlining for sturgeon was conducted in January–March 1975. Subsequently, a 1-month season in May 1975 was established to allow harvest of sturgeon without an impact on salmonids. Setline seasons expanded each year, reached a peak of 9 months in 1982, and were reduced to 7 months in 1983. The months of May-July have been closed to protect spawning sturgeon.

In recent years management and social problems have developed in the setline fishery. Lines are not

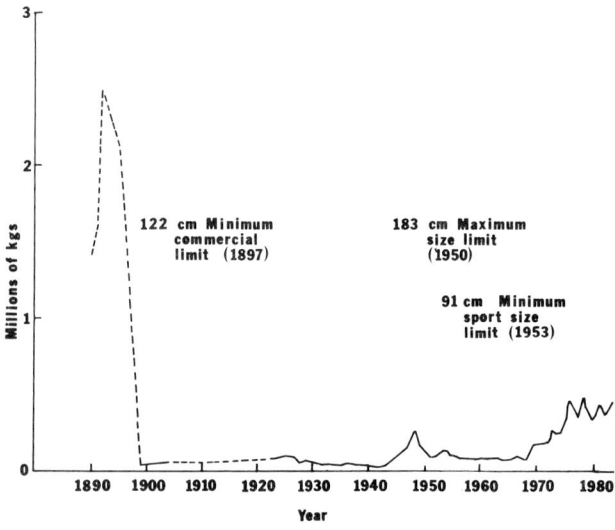

Fig. 1. Columbia River commercial white sturgeon landings, 1898–1983.

Table 1. Commercial and recreational landings of white sturgeon in the Columbia River (1000's of fish), 1970–1983.

Year	Below Bonneville Dam					Above Bonneville Dam					Total Columbia River
	Commercial			Recreational[1]	Total	Commercial			Recreational	Total	
	Gillnet	Setline	Total			Setnet	Setline	Total			
1970	6.3	–	6.3	7.2	13.5	0.4	–	0.4	–	0.4	13.9
1971	7.2	–	7.2	2.8	10.0	0.7	–	0.7	–	0.7	10.7
1972	7.6	–	7.6	5.0	12.6	0.7	–	0.7	–	0.7	13.3
1973	10.7	–	10.7	9.8	20.5	1.1	–	1.1	–	1.1	21.6
1974	10.7	–	10.7	9.9	20.6	0.5	–	0.5	–	0.5	21.1
1975	13.0	1.0	14.0	10.8	24.8	0.6	–	0.6	–	0.6	25.4
1976	18.1	4.7	22.8	15.0	37.8	0.4	0.2	0.6	–	0.8	38.4
1977	8.9	0.8	9.7	22.3	32.0	0.3	0.3	0.7	–	0.7	32.6
1978	8.8	1.0	9.8	29.7	39.5	0.4	0.3	0.7	–	0.7	40.2
1979	18.5	2.0	20.5	30.7	51.2	0.6	0.7	1.3	–	1.3	52.5
1980	6.8	2.6	9.4	25.8	35.2	0.4	1.4	1.8	–	1.8	37.0
1981	10.8	4.1	14.9	25.7	40.6	0.3	1.8	2.1	1.2[2]	3.3	43.9
1982	7.0	4.6	11.6	23.2	34.8	0.2	1.1	1.3	–	1.3	36.1
1983	9.5	2.9	12.4	33.7	46.1	0.3	1.1	1.4	1.8[3]	3.2	49.3

[1] Estimated catches for: 1970–74 March-September, 1975 February-March, May-September, 1976 March, May-October, 1977 March-October, 1978–79 February-November, 1980-83 February-October
[2] Represents estimated catch for June-July only in the area from Bonneville Dam to McNary Dam.
[3] Represents estimated catch for May 25-September 6 only in John Day and McNary Dam tailrace areas only.

checked in a timely manner and some sublegal mortality is caused by snagging. There are general gear problems (loss, theft, vandalism), sport-commercial conflicts (91–122 cm sublegals kept as sport fish, 122–183 cm sport-caught fish sold as commercially caught, and entanglement of sport gear and gill nets with setlines), low return on investment, unrestricted effort, and enforcement problems. It is also difficult to sample a lengthy, low-volume catch season such as the setline fishery. Biologists requested the setline season be reduced to 4 months in 1983, down from 9 months in 1981–1982. However, a 7 month season was set with some tightening of certain regulations to avoid problems. If these problems are not solved, we may be looking at complete closure of the setline fishery in the future.

In May 1982 a test fishery using large, 23 cm minimum mesh gill nets was evaluated as a potential commercial management tool to harvest sturgeon without impacting salmonids. Subsequently, a 5 day season was conducted from January 25–29, 1983, a time frame when salmonids are relatively scarce.

Results from the actual fishery paralleled that of the test fishery, with 40% legal, 60% sublegal, and <1% oversize fish taken. The large-mesh fishery appears to be a viable method for commercial harvest.

Prior to 1975 anglers on the lower Columbia River were mainly interested in salmon (*Oncorhynchus* spp.) and steelhead (*Salmo gairdneri*). Beginning in 1975, with extended salmon and steelhead angling closures, sturgeon angler trips and catches began to climb dramatically. Angler trip totals have established new record highs each year since 1976, numbering 136 000 in 1983. The catch climbed from a low of 2800 fish in 1971 to a peak of 33 700 in 1983 and has averaged nearly 30 000 fish the last 3 years (King 1983b).

Research

Some studies are being conducted to update information gathered during earlier studies. White sturgeon have a relatively long life span, grow very large (579 kg and 675 kg in the Columbia and Snake rivers, respectively [Galbreath 1979]), and are slow to mature. Males mature at about 12 years of age at a length of about 122 cm and females at 15–20 years and 168–183 cm. Of course, older females spawn less frequently with varying intervals between spawning. At least 99.5% of the females taken commercially in the Columbia River are immature and only small amounts of caviar are available. This valuable caviar in oversize fish is quite an inducement for illegal marketing.

Researchers in the past have had little success locating white sturgeon eggs and larvae. We know spawning white sturgeon need water temperatures around 9–15° C. Water temperatures in May are in the middle of the range.

Biologists from Washington and Oregon used an ichthyoplankton net in an attempt to capture white sturgeon eggs and larvae (Stockley 1981). The following results were obtained with net tows in Columbia River flows of 55 000–102 000 cubic meters per second, 12–18° C water temperature, and 3–21 m depth: (1) May–June 1979 – 29 tows yielded 12, 2–3 day-old larvae; (2) April–mid June 1980 – five larvae and several eggs; (3) May 1981 – five larvae.

Extensive sampling conducted in 1983 by WDF during late April–mid June at flows of 76 000–123 500 $m^3 s^{-1}$, 12–17° C water temperature, and 2–15 depth yielded 183 white sturgeon larvae and 48 eggs (Kreitman personal communication).

Over 1300 white sturgeon fin rays were collected from 1980–1983 for age determination. Sturgeon grew at an annual mean rate of 6.6 cm, with a range of 2.1–14.0 cm between the ages of 1–21 years. White sturgeon enter the sport fishery at a minimum of 91 cm and a mean age of 9 years and the commercial fishery at 122 cm and mean age of 12 years. Fish leave both fisheries at a maximum length of 183 cm and a mean age of 20 years (Fig. 2). Comparisons with a similar study (Bajkov 1951) indicated that sturgeon are slightly slower growing now with considerably more age range for each length. This slower growth could be related to a larger stock density with less available food supply, which may be limiting growth (Hess 1984).

A tagging study was conducted by OFC during the late 1940s and early 1950s. A new tagging study was begun in 1965 and has continued with small

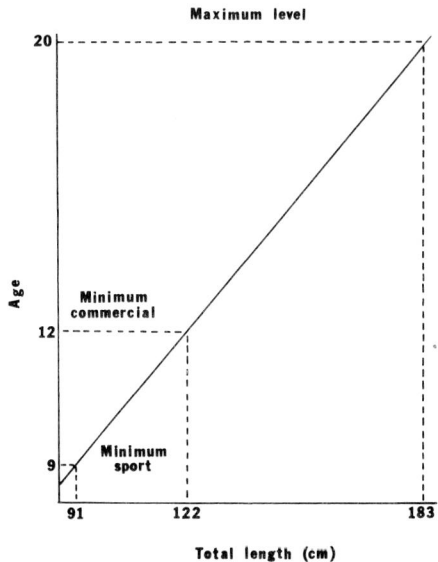

Fig. 2. Length-age relationship of white sturgeon from the lower Columbia River.

numbers of fish tagged each year. Since 1976 more than 1000 white sturgeon have been tagged each year. Through 1983 we have tagged 11 519 white sturgeon in the lower Columbia and have 1141 recoveries (10.1%).

Our purpose in tagging is to determine growth rates and define migrational patterns. Growth data from tag recoveries indicate that sturgeon between 70 and 130 cm TL grow an average of 10 cm per year. To date we cannot define a set migrational pattern. Movement based on tag recoveries appears to be random. Food availability is important, particularly in the winter and early spring when many sturgeon follow the smelt run upriver from the ocean, and in the fall when anchovies are in the estuary. However, this may not be as important as previously thought, as many sturgeon do not exhibit migratory behavior.

Tagging demonstrated that the May 18, 1980, eruption of Mt. St. Helens had a definite effect on sturgeon. The eruption caused a number of sturgeon to leave the Columbia River, particularly on the Washington side. Prior to May 18, 1980, we had 526 tag recoveries of which only 4 (1%) were from river systems other than the Columbia (such as the Oregon's Yaquina River and Washington's Neah Bay in Puget Sound). From May 18, 1980, through 1981, 174 tag recoveries were made with 25 (14%) from other systems. The out-of-system movement was not as pronounced for 1982 with 218 recoveries made; 11 (5%) from outside the Columbia.

Some of the most interesting recoveries were: the fastest growing sturgeon grew 27.9 cm per year; the fastest moving sturgeon went 37 km in 3 days; the farthest recovery was a sturgeon traveling to Puget Sound, WA (a distance of 608 km in 459 days); the longest time out was 8.1 years with a growth of 70 cm, and movement of 35.4 km upstream; and the oddest recovery was from a crab fisherman that found a dead 122 cm sturgeon in a crab pot off the mouth of the Columbia (King in 1983 speech at American Fisheries Society Chapter meeting in Corvallis, OR).

We have problems keeping tags on sturgeon. Nets knock them off or pull them out. We have had the best success with a spaghetti loop tag (overhand knot) inserted at the base of the dorsal fin. The return is about 11%. With other types of tags we have had only a 6% or less return. We also have problems getting some commercial fishermen to report tags. We believe that a high percentage (80%) of tagged fish handled by the commercial fishery may go unreported. Another problem is getting anglers to report tagged sublegals. In 1982, using tagged-untagged ratios, we determined that 81% of the tagged sublegal 20% of the tagged legal fish were going unreported in the sport fishery (King 1983).

Future

What does the future hold for white sturgeon stocks and fisheries below Bonneville? The key to maximum sturgeon production is proper spawning escapement. While measures of spawning escapement are sketchy at this time, we believe the sturgeon stock on the lower Columbia is healthy. The recent 10 year catch level is the highest in this century, suggesting a large stock. Key juvenile indicators are also currently positive, i.e., larval sampling, and the number of sublegal fish handled by the sport and commercial fisheries. However, we are getting some indications that the white sturgeon population may be stabilizing after rapidly increasing in the 1970s. The commercial and sport catches have leveled off in the last several years and the stock may be near a maximum sustainable yield level (Fig. 3).

Table 2 lists the relative catch by size group and indicates stable recruitment in the 152–183 cm length category, implying stable recruitment into the spawning escapement. Commercial fisheries commonly handle oversize sturgeon, and the sport fishery was estimated to have handled 600 and 2200 oversize white sturgeon in 1982 and 1983.

It appears the sport fishery, as it now exists, cannot deplete the sturgeon resource. The sport fishery allows sufficient sturgeon in the 92–122 cm length category to reach the 123–152 cm category, allowing a productive commercial fishery (King personal communication).

Table 2. Estimated landings of white sturgeon (1000's of fish) in legal centimeter-length groups in lower Columbia River commercial and recreational fisheries, 1977–1983.

	91–122 cm Sport	123–152 cm			153–183 cm			Grand total
		Sport	Comm.	Total	Sport	Comm.	Total	
1977	17.4	3.8	9.1	12.9	1.1	0.6	1.7	32.0
1978	22.6	5.5	9.2	14.7	1.6	0.6	2.2	39.5
1979	23.0	6.0	19.2	25.2	1.7	1.3	3.0	51.2
1980	20.4	3.9	9.1	13.0	1.5	0.3	1.8	35.2
1981	20.2	4.2	14.2	18.2	1.3	0.7	2.0	40.6
1982	18.2	4.0	10.8	14.8	1.0	0.8	1.8	34.8
1983	24.6	6.7	11.2	17.9	2.4	1.2	3.6	46.1
1977–83 Avg	20.9	4.9	11.8	16.7	1.5	0.8	2.3	39.9

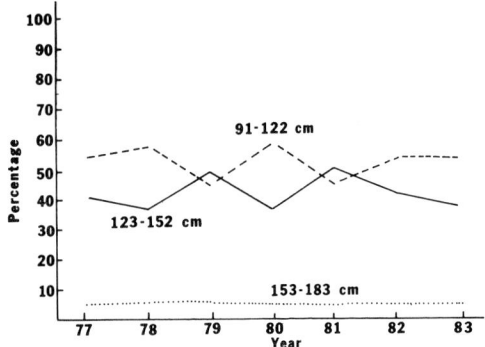

Fig. 3. Relative catch proportion of white sturgeon in centimeter-length groups from sport and commercial fisheries of the lower Columbia River, 1977–1983.

The current commercial fishery is capable of harvesting a good percentage of legal-size fish, but with regulations established to protect the brood stock, we should not see the overexploitation experienced in the 1890s. We will continue to monitor the sturgeon fisheries of the Columbia River. The time frame necessary to recover from a management mistake, such as overfishing, is much longer for sturgeon than for other fish.

Fisheries above Bonneville Dam

The treaty Indian set-net fishery does not catch many sturgeon. Setline landings, although on the increase, do not approach the magnitude of landings by the below-Bonneville Dam fisheries (Table 1). Although sturgeon are found throughout the reservoirs, the non-Indian sport fishery is more concentrated at the bases of various dams. Attempts are now being made to estimate this sport catch.

Research

In 1974, 195 sturgeon were tagged in the Bonneville and John Day pools. To date we have had a 5% recovery rate. This was not a funded project. An Indian sturgeon fishermen employed by ODFW was allowed to keep and sell the legal-size sturgeon (122–183 cm) caught in his setlines and nets. Obviously, we could not attempt a stock estimate. As this was a treaty Indian fishing area and Indians wished to take oversize sturgeon, federal funds were made available for a 3 year study conducted by the U. S. Fish and Wildlife Service (USFWS). Malm (1980) published results of tagging in the Bonneville Pool and Tom Macy (USFWS) intends to publish soon on the John Day pool. The Dalles pool remains to be studied. Funding was not continued as hoped and Malm was unable to delve into the early life history of sturgeon above Bonneville. ODFW and WDF biologists have not been successful in their attempts to find eggs or larvae. Proposals have been prepared to obtain funding for early life history studies. There is little passage from pool to pool over fish ladders and stocks are essentially landlocked. There is some passage through navigation locks and fish were elevated over Bonneville Dam in earlier years (Donaldson 1948).

Malm (1980) tagged 2405 sturgeon and recaptured 161, for a 6% recovery. Abundance was estimated at 31 691 fish using Schnabel's formula with Chapman's modification. The 95% confidence limits were 25 981 and 40 494. This estimate was based upon fish captured during the study which ranged in size from 30.5 cm to 245.1 cm. The catch consisted of 98% sub-legal sport fish (<91 cm TL), 17% legal sport (91–183 cm TL), and 5% oversize (<183 cm. TL).

Average growth was 3.4 cm per year from age 3–12 years, was relatively constant at 16.9 cm per year from age 12–18 years, and was 8.1 cm per year from age 19–28 years. Growth pattern comparisons made with lower Columbia River, Snake River (Idaho), Fraser River (BC), and San Pablo Bay (CA) stocks indicated that sturgeon having free access to ocean and estuary environments grew at a constant rate compared to those landlocked by dams. Malm also stated that a life history management data base (life history data from fish <3 years and >28 years, maturation, timing, and spawning and rearing locations) must be obtained.

We also need to determine the extent of egg resorption in female white sturgeon in the reservoirs, determine the effect of polychlorinated biphenyls (PCBs) on eggs, and answer the question as to whether we have too many large, unproductive white sturgeon for the available food supply.

Conclusions

The catch level of white sturgeon since 1973 remains the highest of the century, with the 1983 catch well over the recent 10 year average. Catches by size group for each fishery indicate stable recruitment in the 168–183 cm category, implying stable recruitment into the spawning escapement. Creel census interviews in 1983 indicated about 2200 oversize white sturgeon were handled compared to 650 in 1982. The recreational catch, angler trips, and average total length of legal size sturgeon were all record highs in 1983. The white sturgeon catch per angler trip in 1983 was greater than 1981 and 1982 (ODFW-WDF reports 1983).

Research conducted in 1983 on spawning characteristics of white sturgeon again documented successful spawning below Bonneville. The number of 1 to 4 year old white sturgeon observed in the commercial shad fishery increased again in 1983. This information suggests the white sturgeon stocks of the lower Columbia River below Bonneville Dam are healthy at the present time. The primary emphasis on problem solving research should be directed above Bonneville Dam.

Acknowledgments

I appreciate the opportunity provided by this symposium to aid in gathering together some of the widely scattered information on sturgeon. Special thanks are due the Oregon Department of Fish and Wildlife for sending me to this symposium and to departmental employees, Steve King for editorial comments, Joanne Hirose for typing and format, and Shirley McKinney for constructing the figures.

References cited

Bajkov, D. 1949. A preliminary report on the Columbia River sturgeon. Oregon Fish Comm., Research Briefs 2: 3–10.

Bajkov, D. 1951. Migration of the white sturgeon in the Columbia River. Oregon Fish Comm., Research Briefs 3: 8–21.

Craig, J.A. & R.L. Hacker. 1940. The history and development of the fisheries of the Columbia River. U. S. Fish Bull. 32: 133–216.

Donaldson, I. 1948. Passage of fish over Bonneville Dam-Col. R. Oregon and Washington. Army Corps of Eng., Portland. 21 pp.

Galbreath, J. L. 1979. Columbia River colossus – the white sturgeon, Oregon Wildl. 34: 3–8.

Hess, S.S. 1984. Age and growth of white sturgeon in the lower Columbia River, 1980–83. Proc. Rept. OR Dept. Fish Wildl., Portland. 15 pp.

King, S. D. 1983. The 1982 Lower Columbia River recreational fisheries – Bonneville to Astoria. OR Dept. Fish Wildl., Portland. 42 pp.

Malm, G. 1980. White sturgeon, *Acipenser transmontanus*, population characteristics in the Bonneville Reservoir of the Columbia River (1976–1978). U. S. Fish Wildl. Serv., Fish. Asst. Off., Vancouver. 28 pp.

Stockley, C. 1981. Columbia River sturgeon. Prog. Rep. 150, Wash. Dept. of Fish., Seattle. 28 pp.

Received 15.5.1984   Accepted 20.3.1985.

Status of white sturgeon, *Acipenser transmontanus*, in Idaho

Timothy G. Cochnauer, James R. Lukens & Fred E. Partridge
Idaho Department of Fish and Game, Lewiston, ID 83501, U.S.A.

Keywords: Abundance, Density estimates, Age, Growth, Management

Synopsis

A five year study was initiated in 1979 to obtain life history information on white sturgeon, *Acipenser transmontanus*, and determine distribution and abundance in the Snake and Kootenai rivers in Idaho. A total of 1266 white sturgeon were sampled from the Snake River and 418 from the Kootenai River. In the Snake River, white sturgeon were most abundant in the sections below Hells Canyon Dam (331 fish sampled) and between Bliss and C.J. Strike dams (935 fish sampled). For the 3 study areas, white sturgeon growth was greatest for fish sampled from the Snake River between Bliss and C.J. Strike dams. Management regulations for all stocks of white sturgeon in Idaho should be continued as catch and release with consideration for possible closure of any fishing for sturgeon in three sections of the Snake River above Brownlee Reservoir because of low numbers.

Introduction

The white sturgeon, *Acipenser transmontanus*, is the only species of sturgeon in Idaho and is found in two major river drainages within the state (Fig. 1). It occurs in 765 km of the Snake River downstream from Shoshone Falls, a geologic barrier to natural distribution upstream. It is also found in two major tributaries of the Snake River, the Clearwater and Salmon rivers. A naturally occurring, isolated stock inhabits the entire Idaho portion of the other major drainage, the Kootenai River in northern Idaho.

Prior to the 1900s, white sturgeon could move freely between the Snake and Columbia rivers, but since that time numerous hydroelectric facilities (8 in Idaho) constructed on the Snake River created barriers to upstream and downstream movements. None of these facilities were constructed with adequate fishways for interchange of white sturgeon between dammed sections. For this reason, isolated stocks of white sturgeon exist between dams on the Snake River. Construction of these impoundments has resulted in a 37% loss of free flowing white sturgeon habitat in the river.

The Kootenai River white sturgeon was isolated from the Columbia River system at the close of the last glacial period in North America. This stock can move freely, within the 190 km of river between Kootenay Lake in Canada through Idaho to Kootenai Falls in Montana. Although barriers to migration have not been constructed in this section of river, extensive land practices in the lower 126 km may have affected adversely or eliminated important white sturgeon habitat. Libby Dam, constructed in 1972 above Kootenai Falls, has changed the natural flow regime in this section by reducing spring flows 50% and increasing winter flows by 300%.

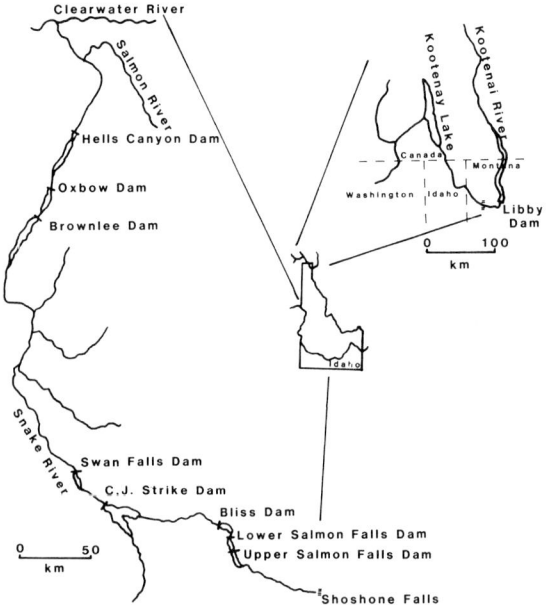

Fig. 1. Natural range of white sturgeon in Idaho.

The first white sturgeon fishing regulations in Idaho were adopted in 1943 when commercial fishing was prohibited and possession limits of two per day were imposed. A yearly limit of 2 white sturgeon was implemented in 1956. Catch and release fishing regulations for white sturgeon were established in 1970 on the Snake River and in 1984 on the Kootenai River.

In 1979 the Idaho Department of Fish and Game initiated a 5 year research program to assess the status of white sturgeon within the state and collect life history information. The Department's long term white sturgeon management plans include the continuance of catch and release fishing until such time as information on the species supports the re-establishment of consumptive fishing.

In our studies we sampled white sturgeon within the entire Idaho portion of the Kootenai River, the Snake River below Hells Canyon Dam and the Snake River above Brownlee Dam. The river section between Brownlee and Hells Canyon dams was not included.

Methods

Abundance

Relative abundances of white sturgeon in the study areas were assessed using rod and reel and multiple hook set lines. Attempts to capture sturgeon using nets, traps and electrofishing gear were unsuccessful. Rod and reel terminal gear included a single 1/0 to 5/0 hook attached to 18 kg test monofilament line. Set lines had six single hook droppers spaced a meter apart, and were attached to a 10 m main line with harness snaps. Each set line was weighted at each end and marked by a float line tied to one end. 'Nightcrawler' worms or fresh fish were used as bait for both rod and reel and set lines.

Fishing sites were primarily limited to pools in excess of 8 m in depth. We generally fished 3 to 4 days per week throughout the year, depending on weather conditions. The study in Hells Canyon was conducted only during the months of April through October. Set lines were fished for 24 h and checked every 4 h during daylight hours. Rod and reel fishing occurred from sunup to sundown. The number of hours fished for each gear type was tabulated daily. Generally four rods at a time were fished from the bank. Set lines were set and checked using river sleds.

For individual recognition when recaptured, each white sturgeon was marked with a three-digit tattoo on the ventral portion of the rostrum, immediately anterior to the mouth with a commercial horse lip tattoo gun and black tattoo ink. White sturgeon larger than 125 cm were also marked with plastic 15 cm Floy anchor or spaghetti tags at the dorsal fin insertion.

In river sections where the numbers of sturgeon collected were adequately large (>100), density estimates were calculated using a multiple census, mark-recapture method (Schnabel 1938). Calculations for estimates were begun 1 year after initiation of a study on a particular river section. Recaptures occuring in the same time interval as initial capture were not included in the calculations for that time interval. Each time interval ($\hat{t}$) was a 2 month period. The following formula was used: $1/\hat{N} = (m_{\hat{t}} r_{\hat{t}})/(c_{\hat{t}} m_{\hat{t}}^2)$. Confidence intervals (95%)

were computed using: Variance of $1/\hat{N} = S^2/c_{\dagger}m_{\dagger}$, $S\,1/\hat{N} = \sqrt{Var\,1/\hat{N}}$. Using † distribution and †−1 degrees of freedom, $1/\hat{N} + (\dagger \times (S\,1/\hat{N})) =$ confidence interval, where, $\hat{N}$ = density estimate, $c_{\dagger}$ = total sample taken during sampling interval, $m_{\dagger}$ = total marked fish at large at start of sampling interval, $r_{\dagger}$ = number of recaptures in sampling interval, and $S^2 = ((r_{\dagger}^2/c_{\dagger}) - (r_{\dagger}m_{\dagger})^2/(c_{\dagger}m_{\dagger}^2))/(\dagger-1)$.

Age and growth

White sturgeon total lengths were measured to the nearest 0.5 cm. Weights were determined to the nearest 0.1 kg. Captured sturgeon were categorized into three length groups for comparisons; less than 91.5 cm total length, 91.5 to 183.0 cm, and greater than 183.0 cm. Sturgeon between 91.5 and 183.0 cm were within the legal harvest range for the Snake River prior to 1970 and the Kootenai River prior to 1984.

A section of the first pectoral fin ray was removed from each white sturgeon captured for age determination following techniques described by Cuerrier (1951) and Coon et al. (1977). The fin ray sections (5 to 25 mm in width) were removed 5 to 25 mm from the fin articulation using side-cutters or a fine-toothed hacksaw blade. The sections were polished with 400–600 grit sandpaper and read using a binocular dissection microscope under reflected light. Annuli were visible as light bands alternating between dark bands representing growth periods.

Length-weight relationships are presented as a power function $W = aL^b$, where W = body weight in kg, L = total length in cm, and a and b are parameters. The parameters a and b were estimated by taking logarithms (base 10) of both sides of the equation, $\log W = \log a + b \log L$. Log a and b were estimated for pairs of lengths and weights by GM regression techniques.

Results

Comparisons of sturgeon abundance between river sections were not possible because of the different sampling seasons for the sections. However in the Snake River above Brownlee Reservoir, sampling was conducted during comparable times. Considerably more white sturgeon were captured in the section between Bliss and C.J. Strike dams during the three year study (Table 1). Two percent of the white sturgeon captured in this section were greater than 183 cm; more (numerically) than any other river section studied. In the other three sections studied above Brownlee Reservoir, there were white sturgeon absent in at least one of the three designated length categories.

Density estimates in the three study areas in which over 100 white sturgeon were caught show that white sturgeon were most abundant in terms of fish per kilometer in the Snake River below Hells Canyon Dam (Table 2).

Age and growth

White sturgeon sampled in all river section ranged in age from 2 to 53 years (Fig 2). White sturgeon were fully recruited to the fishing gear type at age 4. We captured more fish over the age of 20 in the Kootenai River than in any Snake River section. The modal age of white sturgeon captured in the Kootenai River was 19 as compared to 4 and 5 in the Snake River.

Most of the white sturgeon captured in the Kootenai River were within the length group, 91.5–183.0 cm. The modal length of those fish was 115 cm while in the Snake River modal lengths were 60 and 70 cm (Fig 3).

White sturgeon from the Snake River above Brownlee Dam were larger at all ages than white sturgeon from either Hells Canyon or the Kootenai River (Fig. 4). Growths of white sturgeon from the Kootenai River and the Hells Canyon area were relatively similar until just past 20 years of age where the two growth curves began to diverge. White sturgeon older than 22 years from the Kootenai River were smaller than similar-age fish from either of the two Snake River sections.

The length-weight relationships for these three white sturgeon stocks were: (1) Snake River-above Brownlee Dam, $W = 3.0 \times 10^{-7} L^{3.61}$ (N = 560); (2) Snake River-below Hells Canyon Dam, $W = 7.76 \times 10^{-7} L^{3.43}$ (N = 305); and (3) Kootenai River,

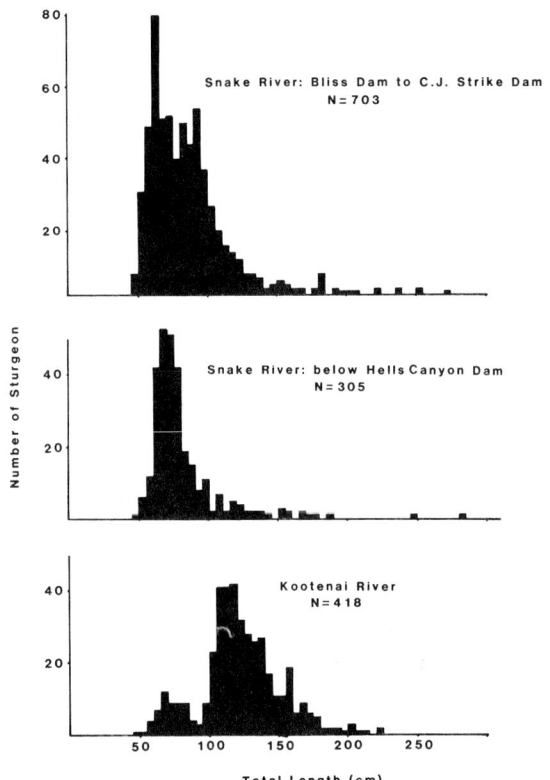

Fig. 2. Length frequency of white sturgeon sampled from the Snake River below Hells Canyon Dam and between Bliss and C.J. Strike dams and from the Kootenai River, 1979–1983.

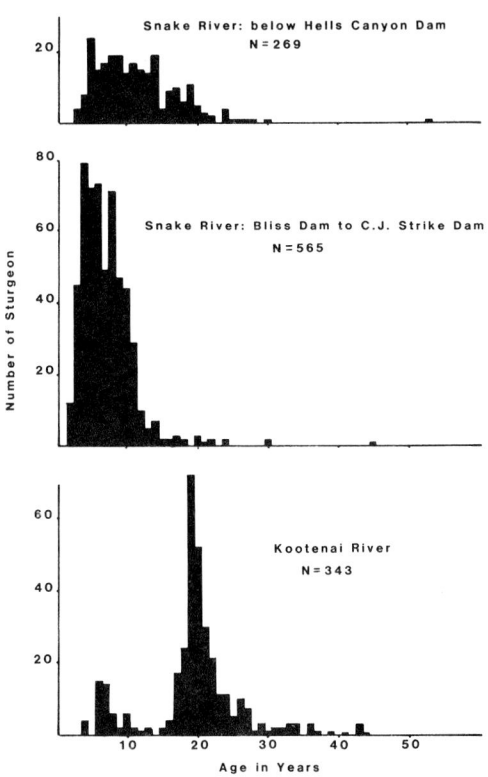

Fig. 3. Age frequency of white sturgeon sampled from the Snake River below Hells Canyon Dam, between Bliss and C.J. Strike dams and from the Kootenai River, 1979–1983.

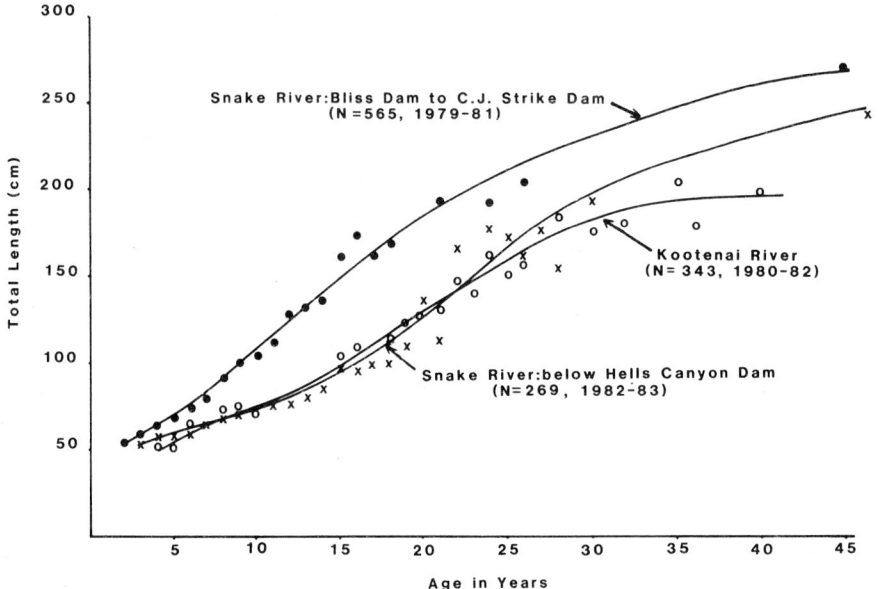

Fig. 4. Age length relationships for three white sturgeon stocks sampled in Idaho, 1979–1983.

Table 1. Numbers of white sturgeon captured, catch per effort and length group composition of white sturgeon caught and released from the Snake and Kootenai rivers, Idaho, 1979–83.

River section	Total number of sturgeon captured	Percent recaptures	Catch per 1000 rod hours	Percent in length groups		
				<91.5	91.5–183.0	>183.0
Snake River –						
Shoshone Falls to Bliss Dam	20	45.0	0.9[a]	0	89	11
Bliss Dam to C.J. Strike Dam	905	26.0	54.5[a]	68	30	2
C.J. Strike Dam to Swan Falls Dam	9	0	7.2[a]	89	0	0
Swan Falls Dam to Brownlee Reservoir	1	0	1.8[a]	100	0	0
Hells Canyon Dam to Lower Granite Pool	331	7.6	228.8[b]	83	15	2
Kootenai River –						
Idaho border to Canada border	418	15.1	27.7[a]	13	84	3

[a] Year around
[b] April–October

Table 2. Estimated abundance of white sturgeon in the Snake and Kootenai Rivers, Idaho, 1979–83.

River section	Section length (km)	Estimated abundance	95% Confidence interval	Estimated white sturgeon per km
Snake River–				
Bliss Dam to C.J. Stike Dam	86.3	2192	1479–4276	25.4
Hells Canyon Dam to Lower Granite Pool	74.0	2785	1472–4733	37.6
Kootenai River –				
Bonners Ferry to Canada border	76.0	1148	907–1503	15.1

$W = 1.66 \times 10^{-6} L^{3.20}$ (N = 341) [W = body weight in kg, L = total length in cm, N = sample size].

Discussion

Sturgeon captured upstream of Upper Salmon Falls Dam were longer for any given age than other white sturgeon aged in Idaho. These fish have apparently failed to form annuli and are probably older than the rings on their pectoral fins indicated. Large springs in this river section provide areas of warm constant temperatures and sufficient food so that growth could occur on a year around basis thereby allowing for greater growth as well as providing a situation where recognizable annuli would not be formed.

When growth is compared between three stocks of white sturgeon in Idaho, sturgeon from the Snake River section above Brownlee Reservoir, were larger at all ages than sturgeon from the Hells Canyon section or the Kootenai River. Relative growth of white sturgeon appears to be related to water temperatures. Faster growing white sturgeon in the upper Snake River sections may be a result of relatively higher year-around water temperatures resulting from the large volume of spring water entering the area.

Large numbers of white sturgeon in the Snake River were found only between Bliss and C.J. Strike dams and below Hells Canyon Dam. Both of these sections presently support catch and release white sturgeon fisheries. In other sections of the Snake River, the stocks may not be of adequate numbers to support any fishery if associated hooking mortality or illegal harvest result in further decline of numbers.

Upstream from Bliss Dam, the Snake River supports some white sturgeon but recruitment appears to be depressed. Successful reproduction may not have occurred within this section since 1974, as that was the youngest year class of sturgeon observed. The lack of recruitment is probably not a result of a lack of suitable habitat, but rather the result of low spawner numbers coupled with low spawning frequency.

Female white sturgeon may not mature until at least 11 years of age and probably do not spawn but every 3 to 11 years (Table 3). Eleven year old white sturgeon in this section of the Snake River were approximately 125 cm in length. Only two fish older than 11 years and six fish greater than 125 cm were observed above Bliss Dam.

The Snake River between Swan Falls and C.J. Strike dams supports fewer white sturgeon than the section immediately upstream. The presence of young fish (5 years of age) was evidence of successful recent spawning. However, assumed sturgeon spawning habitat was limited to a short distance immediately below C.J. Strike Dam and may be the factor limiting population size in this section. Scott & Crossman (1973) describe white sturgeon spawning habitat as areas of hard substrate with moderate current.

Below Swan Falls Dam, the Snake River resembles the 86 km river section below Bliss Dam and would be expected to support similar numbers of white sturgeon. The catch of only one sturgeon from this section indicates that the stock is depressed and has been reduced since the early 1970s. With an expenditure of rod and reel effort in 1972 comparable to this study, Idaho Department of Fish and Game personnel captured and released 22 white sturgeon from this river section. These fish ranged in total length from 60 cm to over 183 cm.

Physical habitat and water quality below Swan Falls Dam have not noticeably changed since that time, so other factors may be involved in the density decline. Sport fishing in this river section has increased at least three-fold since the early 1970s as a result of the establishment of a channel catfish, *Ictalurus punctatus*, fishery from releases initiated in the 1960s. Increased fishing for channel catfish may have increased illegal harvest and mortality of white sturgeon in this section because of the similarities in fishing techniques and locations for both species.

Both relatively abundant sturgeon stocks in the Snake River (below Hells Canyon Dam and between Bliss and C.J. Strike dams) have apparently responded favorably to catch and release regulations since 1970. Juvenile recruitment in Hells Canyon appears adequate and in similar proportion to that found in this section in the early 1970's by Coon (1977). Sturgeon less than 91.5 cm comprised 86% of the sample in 1972–75 as compared with 82% in 1982–83. It appears that the length group between 91.5 and 183.0 cm (the length group harvested prior to 1970) has somewhat recovered since the establishment of catch and release regulations. It has required approximately 10 years of protection for this length group to grow to 15% of the stock sampled from the 3% observed during the 1972–75 study.

Between Bliss and C.J. Strike dams, the length group of sturgeon between 91.5 and 183.0 cm comprised 30% of the total stock sampled. This would be a ten-fold increase over levels found in Hells Canyon during the early 1970s, assuming that harvest impacted both stocks similarly.

The white sturgeon stock in the Kootenai River in Idaho appears to be comprised primarily of mid-

Table 3. Age in years of first maturity and spawning frequency (years) of white sturgeon reported from various studies.

Author	Male	Female	Male	Female	Location
Pycha (1956)	–	11–12	–	–	Sacramento River, CA.
Semakula & Larkin (1968)	11–12	11–34	4–11	6–8	Fraser River, B.C.
Galbreath (1979)	–	–	–	2–11	Columbia River, WA.
Kohlhorst (1980)	–	14	–	–	Sacramento River, CA.
Stockley (1981)	–	15	–	3–5	Columbia River, WA.

size and larger fish. Only 12.9% of the fish sampled were less than 91.5 cm in length, compared to over 80% for the two abundant Snake River stocks. Assuming that young aged white sturgeon could be caught with similar expenditures of effort in both the Snake and Kootenai rivers, we would not expect to see missing age classes between 2 and 15 years. It appears that sturgeon recruitment has been decreasing since the mid-1960s. The causes of this decline may be a combination of environmental changes in the river system including the loss of sloughs and marshes along the river because of dike construction, increases in levels of chemical pollutants from upstream sources, and changes in the flow regime downstream of Libby Dam.

We found no evidence that white sturgeon move around dams. This suggests each white sturgeon stock is isolated from other stocks either upstream or downstream of a dam, and that each isolated stock can be managed independently of others.

These recent studies on white sturgeon stocks in Idaho have provided us with baseline information for management direction and comparison with future studies. Two Snake River stocks have apparently responded well to catch and release fishing and should be managed as such in the future. The other sections of the Snake River do not look as favorable, and probably should be closed to catch and release white sturgeon fishing.

In response to the limited juvenile recruitment to the Kootenai River white sturgeon stock, consumptive fishing was terminated after 1983 although a catch and release fishery is still allowed. The Kootenai River should be continued under catch and release white sturgeon fishing until future studies either support a complete closure on fishing for sturgeon or data suggests the stock can survive consumptive fishing with the present structure of the stock.

References cited

Coon, J.C. 1977. Abundance, growth, distribution, and movements of white sturgeon in the mid-Snake River. PhD Dissertation, University of Idaho, Moscow. 63 pp.

Cuerrier, J. 1951. The use of pectoral fin rays for determining age of sturgeon and other species of fish. Can. Fish Cult. 11: 10–18.

Galbreath, J. 1979. Columbia River colossus-the white sturgeon. Oregon Wildlife, March 1979: 3–8.

Kohlhorst, D.W. 1980. Recent trends in the white sturgeon population in California's Sacramento-San Joaquin estuary. California Fish and Game 66: 210–219.

Pycha, R.L. 1956. Progress report on white sturgeon studies. California Fish and Game 42: 23–35.

Schnabel, E.Z. 1938. The estimation of total fish population of a lake. Amer. Mathem. 45: 348–352.

Semakula, S.N. & P.A. Larkin. 1968. Age, growth, food and yield of the white sturgeon, *Acipencer transmontanus*, of the Fraser River, British Columbia. J. Fish. Res. Board Can. 25: 2589–2602.

Scott, W.B. & E.J. Crossman. 1973. Freshwater fishes of Canada. Fish. Res. Board Can. Bull. 184.966 pp.

Stockley, C. 1981. Columbia River sturgeon. State of Washington, Department of Fisheries, Progress Report 150.28 pp.

Received 15.5.1984   Accepted 27.2.1985

Management of the lake sturgeon, *Acipenser fulvescens*, population in the Lake Winnebago system, Wisconsin

Daniel J. Folz[1] & Lee S. Meyers[2]
[1] *Wisconsin Department of Natural Resources, Box 2565, Oshkosh, WI 54903, U.S.A.*
[2] *Wisconsin Department of Natural Resources, Box 10448, Green Bay, WI 54307, U.S.A.*

Keywords: Harvest, Exploitation, Mortality, Recruitment, Population estimate, Spawning population, Age, Length-weight

Synopsis

Current management of the lake sturgeon, *Acipenser fulvescens*, by the Wisconsin Department of Natural Resources is centered on limiting the harvest to maintain a sustained yield. To regulate harvest it is necessary to monitor trends in the population of these long-lived fish. From 1975 through 1983, 3380 lake sturgeon were marked at spawning sites. Netting operations in Lake Winnebago marked 2826 lake sturgeon. Lake sturgeon spawn in the Winnebago system from mid April to early May. Placement of rock riprap on the outside bends of tributary rivers has created additional spawning sites. Mean total lengths of male (133.4 cm) and female (160.3 cm) lake sturgeon netted at spawning sites were consistent from 1975 through 1983. From 1955 to 1983, 18075 lake sturgeon were harvested from Lake Winnebago, ranging from 8 to 2238 fish per season. After an increase in the minimum length limit in 1974, the mean harvest increased from 584 to 698. This appears to be the result of increased fishing effort (32.7%) combined with higher population levels. Density estimates from 1976 to 1982 indicate 24 600 lake sturgeon over 114 cm compared to 10 200 from 1955 through 1959. Total annual mortality was 10.5% in recent years compared to 5.4% in the past, however the spearer exploitation rate decreased from 4.3 to 2.5%. Density estimates and catch statistics indicate an increase in the population. The change in mortality is probably the result of increasing recruitment.

Introduction

No sport fishery for lake sturgeon in the United States or Canada is known to surpass that now in existence on Lake Winnebago. Although the spearing season was initiated in 1932, the method was historically practiced by Indians living in the region (Probst & Cooper 1955). Later, as overexploitation became a potential threat to the fishery, the Wisconsin Conservation Department began a research orientated management program. From 1955 to 1967, an extensive study was conducted on the life history along with recruitment and harvest levels of the Winnebago population reported by Priegel & Wirth (1974, 1975).

Current management consists of determining the harvest and size of the stock so that exploitation can be regulated. Overexploitation of a long-lived species is a problem that can take many years to correct. Monitoring the harvest, spawning migrations and marking lake sturgeon during netting operations on Lake Winnebago is necessary to determine if population levels are being maintained and to suggest corrective action, if needed.

The lake sturgeon is classified as rare over much of its original range by the U.S. Fish and Wildlife Service, and the Wisconsin Department of Natural Resources presently has the lake sturgeon on its

Watch Species List. Due to these classifications, it is important that the levels of this viable population be closely monitored.

Description of study area

Lake Winnebago is a large (55 766 ha), shallow, eutrophic lake in east-central Wisconsin (Fig. 1). It has a maximum depth of 6.4 m and an average depth of 4.7 m (Wirth 1959). Lake Winnebago has a methyl-orange alkalinity of 119 to 124 ppm, and pH varying from 7.7 to 8.5. It is characterized by heavy algae blooms in the summer months and in winters with little snow cover. Seventy-six fish species have been reported from Lake Winnebago (Priegel 1967). Two large river systems enter the lake: the 171 km Fox River and the 346 km Wolf River which join in the 11116 ha upriver lakes (Lake Poygan, Butte des Morts and Winneconne). The watershed covers approximately 1 578 974 ha.

History of fishery

A winter spearing season is the only legal means to harvest lake sturgeon in the Winnebago system, although there is a fall hook and line season in other river systems of Wisconsin. Spearing is a method of harvest which has been a tradition on Lake Winnebago. Lake sturgeon spearing occurs in shanties placed over 1 by 1.8 m holes cut in the ice. Shanty counts during the season have not exceeded 3000. The combined area of 3000 spearing holes is approximately 0.5 ha or less than 0.001% of the surface area of Lake Winnebago.

Prior to 1915 there were no restrictions on the harvest of lake sturgeon. Due to concern over the drop in levels of many lake sturgeon populations, the harvest of lake sturgeon was prohibited from 1915 to 1931. The spearing season was established on Lake Winnebago in 1932.

At first, regulations were liberal, allowing spearers to take 5 lake sturgeon per season at a minimum length of 76 cm. In 1954, the season bag limit was reduced to 3 fish, then 2 fish in 1956 and 1 fish in 1958. The minimum length limit was increased to 102 cm in 1955 and 114 cm in 1974. The present season runs from the second Saturday in February through March 1 each year on Lake Winnebago.

The upriver lakes support a smaller lake sturgeon stock which had been over-harvested in the 1950's. Presently, a two day season occurs every fifth year on these lakes. When the stock has recovered, spearing seasons will occur more often. Data on the fishery was described by Priegel & Wirth (1978).

A mandatory registration system was established in 1955, whereby a speared lake sturgeon must be registered at one of the authorized stations along the lake shore by 6:00 p.m. of the same day. A license sales was established in 1960 and since 1980 the license must be purchased prior to the season.

Methods

As each lake sturgeon was registered, total length, weight, location, date and tag numbers of marked fish were recorded. A t-test of means was used to compare changes in harvest levels, lengths and spearer numbers. A length–weight relationship was calculated by the least squares method using fish grouped by 2.54 cm intervals from 114 to 191 cm. An analysis of covariance was run on the length–weight relationship of lake sturgeon harvested during two periods from 1974 to 1983 and 1955 to 1967. During the 1975, 1976 and 1981 seasons, the first pectoral fin ray was taken from 557 lake sturgeon. Dried fin rays were aged following methods of Priegel & Wirth (1975). An estimate of total annual mortality was calculated from a catch curve (Ricker 1975).

During the openwater periods (April to November) each year, a state crew used Lake Erie type trap nets and otter trawls in Lake Winnebago for freshwater drum removal (Priegel 1971). In conjunction with this operation, all legal sized (over 114 cm) lake sturgeon were tagged with monel self-piercing tags (cattle ear type, size 681 and 62) applied at the base of the dorsal fin. Total length, tag number, area of lake and date were recorded for each fish.

Density estimates were based upon lake sturgeon marked in Lake Winnebago and re-

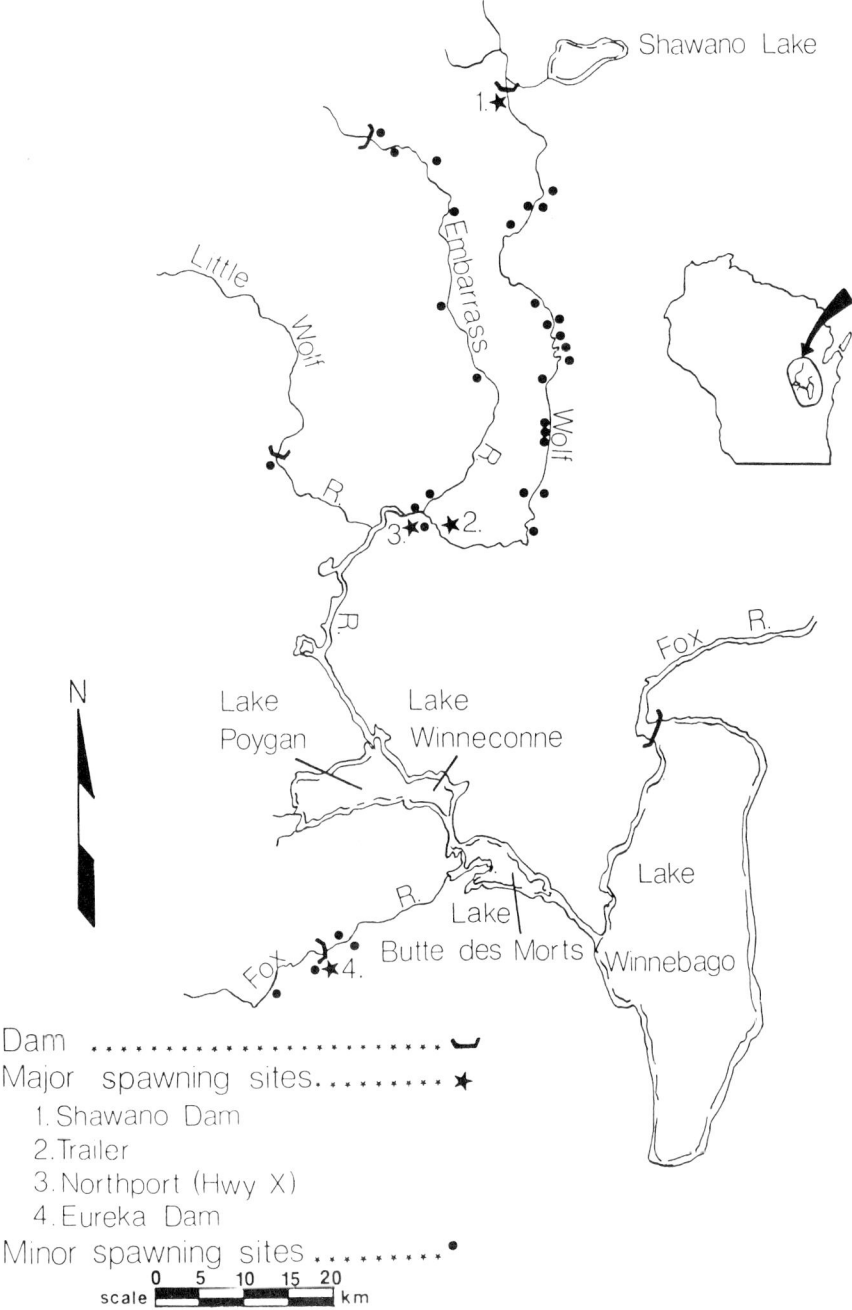

Fig. 1. Map of the Lake Winnebago system, Wisconsin, showing major and minor lake sturgeon spawning sites.

covered through the registration system during the following spearing season. Estimates were calculated from the Peterson mark-recapture formula as modified by Bailey (Ricker 1975). A minimum annual spearer exploitation was determined as the percentage of tags returned from the number available from the tagging effort in the previous year. Rates were considered minimum as non-return of tags by spearers and tag loss were not calculated. News releases, a ten dollar incentive system and spot checks of registration stations were utilized to assure maximum return of tags from spearers.

From 1975 to 1983, lake sturgeon were netted while spawning on rock substrate sites in tributary rivers of Lake Winnebago. Spawning sites were documented 40 km upstream on the Fox River and 200 km upstream on the Wolf River from Lake Winnebago. The first dam on each tributary acts as a barrier except the low head dam on the Fox River during high water years.

Major spawning sites were considered where netting efforts resulted in the capture of more than 20 fish. Minor sites usually resulted in the capture of less than 20 fish and were not utilized by spawn-

ing lake sturgeon each year. Lake sturgeon were captured by handheld dip nets and measured, sexed, tagged and released at the spawning site. All lake sturgeon were handled on canvas tarps to prevent injury. Monel tags (size 49) were locked to the base of the dorsal fin.

Results

A total of 18 075 lake sturgeon were legally harvested from Lake Winnebago from 1955 to 1983. The annual harvest has ranged from 8 fish taken during two seasons (1969 & 1973) to 2238 fish in

Table 1. Lake sturgeon spearing harvest and license sales, Lake Winnebago, 1955–1983.

Year	Season length in days	Number registered	License sales	Percent successful spearers
1955	29	1505		
1956	20	661		
1957	19	851		
1958	22	464		
1959	19	221		
1960	20	520	2688	19.3
1961	19	340	3352	10.1
1962	20	262	2322*	11.3
1963	21	1001	4522	22.1
1964	23	685	5400	12.7
1965	17	718	5727*	12.5
1966	25	300	4285	7.0
1967	26	1424	6014	23.7
1968	28	21	4832*	0.4
1969	22	8	2154	0.4
1970	23	692	4849	14.3
1971	24	159	2396*	6.6
1972	26	1251	5632	22.2
1973	27	8	1739	0.5
1974	21	117	3250	3.6
1975	22	530	4319	12.3
1976	17	936	4528*	20.7
1977	18	287	5454	5.3
1978	19	1246	6891	18.1
1979	20	421	2993	14.1
1980	22	763	5040	15.1
1981	16	407	7673*	5.3
1982	17	2238	6821	32.8
1983	18	39	6000	0.7
Total		18075	108881	13.2

* Upriver lakes season (adjusted license sales; removed estimated 936 extra licenses sold for upriver season)

1982. Extreme fluctuations in the annual spearing harvest of lake sturgeon from Lake Winnebago have been characteristic of the fishery over 29 seasons (Table 1).

Spearing success and harvest rely heavily on water clarity and travel conditions on the ice as reported by Priegel & Wirth (1975). Lake sturgeon were usually speared near the bottom in water depths of 3.7 to 6.1 m, which covers 80% of Lake Winnebago. During some seasons, the water clarity has ranged 1.2 to 3.4 m. This poor clarity was usually caused by an algae bloom if snow cover was less than 30 cm or turbid runoff water from a snow melt. Clarity less than 3.7 m normally resulted in a below average harvest. Spearers became more mobile through the use of 4-wheel drive vehicles and snowmobiles. However, a severe snow storm or excessive thaw during the season restricted spearing to local near shore areas. When travel was unrestricted, spearers were able to move shanties to more productive areas of the lake. If water clarity was excellent allowing visibility to the bottom in 5.5 m and travel was unrestricted throughout the season, the harvest was usually over 1000 lake sturgeon.

In 1974, the minimum length limit was increased from 102 to 114 cm in order to reduce the annual harvest by 11% and bring it in line with the estimated annual recruitment of 540 lake sturgeon (Priegel & Wirth 1975). Mean annual harvest from 1955 to 1973 was 584 (standard error (SE) = 106). From 1974 to 1983, the mean annual harvest was 698 (SE = 207). This was a 19.5% increase in the annual harvest. Due to the high variation in the harvest, there was no significant difference in the means (p<0.05). The 540 level falls within the 95% confidence limits of the mean harvest since the regulation change in 1974 (229≤698≤1167).

From 1960 to 1973, the mean number of spearers per season was 3994 (SE = 402). From 1974 to 1983, a mean of 5297 (SE = 496) spearers participated in the sport. This was a 32.7% increase in pressure. A test of the means showed no significant difference in the two periods (p<0.05). Spearer success rate each season ranged from 0.4 to 32.8% between 1960 and 1983. Overall success was 13.2% (1 lake sturgeon per 7.6 spearers). Although the number of spearers increased by 32.7% (1304 spearers) during the 1974 to 1983 period, the mean success rate remained at 13.2%.

From 1974 to 1983, season length varied between 16 and 22 days and the mean daily harvest was 37 lake sturgeon, compared to 26 fish per day from 1955 to 1973 (Table 1). Over the 10 seasons, the lowest daily production of 2 fish occurred in 1983, while the 17 day 1982 season had a mean daily harvest of 132 fish. The 2238 lake sturgeon taken in 1982 was the highest total and daily catch over 29 seasons of mandatory registration. The second highest season harvest of 1505 fish occurred in 1955 while the 1246 lake sturgeon registered in the 19 day 1978 season resulted in the second highest daily production (66 fish per day).

From 1974 to 1983, there was no trend of increase or decrease in the mean length (140 cm) of lake sturgeon harvested from Lake Winnebago (Table 2). The mean length ranged from 138.7 cm (SE = 0.36) during a high harvest season in 1982 (2238 fish) to a significantly higher length of 145.9 cm (SE = 3.15) during a low harvest season in 1983 (39 fish). This change in length was probably due to other factors than change in the stock structure. Due to the regulation change in 1974 which eliminated the harvest of lake sturgeon in the 102 to 113 cm range, there was a shift in length frequency of the harvest (Fig. 2). There was an increase in the frequency of 114 to 150 cm fish, however the harvest of lake sturgeon over 150 cm remained consistent during the 29 year period.

From 1974 to 1983, the total weight of the harvest was 142 473 kg. Mean weight per fish ranged from 18.1 to 22.6 kg with no trend of increase or decrease (Table 2). Since 1974, 1.3% of the lake sturgeon exceeded 45.4 kg which was similar to past records.

The length-weight relationship for 6984 lake sturgeon registered from 1974 to 1983 was:

$$\text{Log}_{10} W = -5.9053 + 3.3382 \, \text{Log}_{10} L,$$

where $\text{Log}_{10} L$ was total length in centimeters and $\text{Log}_{10} W$ was weight in kilograms. The correlation coefficient (r^2) was 0.99. An analysis of covariance for the 1974 to 1983 period compared to the 1955 to 1967 period indicated no significant difference in

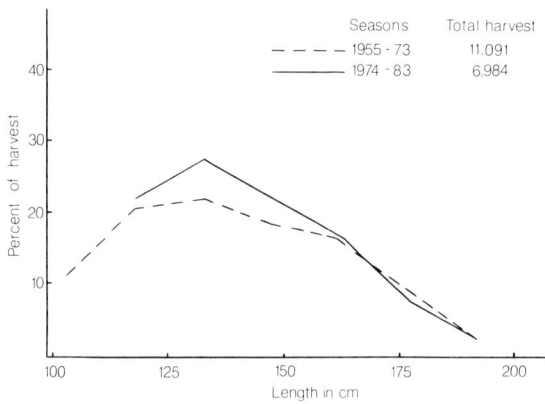

Fig. 2. Length frequency of lake sturgeon harvest, Lake Winnebago, Wisconsin, 1955–1983.

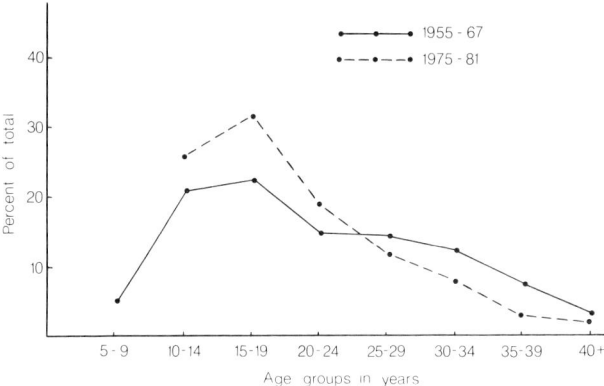

Fig. 3. Age frequency of lake sturgeon harvest, Lake Winnebago, Wisconsin, 1955–1967 and 1975–1981.

weight-length relationships (p>0.05).

Lake sturgeon aged from the 1975, 1976 and 1981 spearing harvests of the Lake Winnebago stock ranged from 10 to 55 years (Table 3). Mean age of the 557 fish was 20.0 years (SE = 0.3) which was a significant decrease (p<0.05) from the mean of 21.5 years for 7013 lake sturgeon harvested from 1955 to 1967. Age frequency of the lake sturgeon indicated a 13% shift from fish over 25 years to fish under 25 years old (Fig. 3) than reported for the 1955 to 1967 period by Priegel & Wirth (1975).

Mean length and weight for most ages in the 1975 to 1981 sample was slightly higher than those reported for lake sturgeon harvested from 1955 to 1967 (Priegel & Wirth 1975). However, lake sturgeon growth was quite variable, for example age 15 fish ranged from 114 to 150 cm. There was a range of lengths in most age classes of 30 cm or more.

Lake sturgeon were fully recruited to the harvest by age 16 and few remained after age 36. Total mortality based upon 21 year classes (age 16 to 36) for 1975, 1976 and 1981 was 6.8, 11.1, and 8.8%, respectively. Total mortality for the three years combined was 10.5% compared to 5.4% from 1955 to 1967 (Priegel & Wirth 1975).

From 1975 to 1982, 2826 lake sturgeon were marked during the open water months in Lake Winnebago. Over the 8 year period, mean length decreased while the number captured each year increased (Table 4). A mean of 245 fish averaging 136.5 cm were captured from 1975 to 1979 com-

Table 2. Mean length (cm) and weight (kg) of lake sturgeon registered, Lake Winnebago, 1974–1983.

Year	Number registered	Mean length	SE	Mean weight	SE
1974	117	140.9	1.5	20.5	0.8
1975	530	140.6	0.7	20.8	0.4
1976	936	138.8	0.5	20.2	0.3
1977	287	142.3	1.0	21.1	0.5
1978	1246	141.4	0.5	21.0	0.3
1979	421	143.0	0.9	21.7	0.4
1980	763	141.1	0.6	20.6	0.3
1981	407	140.2	0.8	18.1	0.4
1982	2238	138.7	0.4	18.9	0.2
1983	39	145.9	3.2	22.6	1.8
Mean	698	140.3		20.4	

Table 3. Age frequency, mean length (cm) and weight (kg) of 557 lake sturgeon harvested from Lake Winnebago, 1975, 1976 and 1981.

Age	Percent total	Mean length	Mean weight	Age	Percent total	Mean length	Mean weight
10	0.5	114.3	8.6	26	3.1	157.2	29.9
11	5.2	118.4	10.9	27	1.6	158.5	32.2
12	5.0	120.9	11.2	28	2.9	158.8	27.7
13	6.5	121.9	11.6	29	1.4	153.2	27.6
14	8.6	125.0	13.3	30	2.2	166.4	34.7
15	6.3	131.6	15.7	31	1.8	163.8	32.6
16	8.4	133.4	15.6	32	1.8	161.0	31.9
17	6.1	136.7	17.1	33	0.9	163.6	34.8
18	6.8	137.9	18.1	34	1.3	164.3	35.1
19	4.0	141.2	19.2	35	1.1	172.7	35.8
20	4.9	140.7	20.0	36	0.4	185.4	47.9
21	3.1	146.1	22.5	37	0.4	172.7	34.5
22	5.2	147.1	23.1	38	–		
23	3.8	152.7	25.7	39	0.7	182.9	46.7
24	1.6	147.1	22.5	40 to			
25	2.9	158.8	29.6	55	1.8	180.1	43.0

pared to a mean of 533 fish averaging 131.1 cm from 1980 to 1982. The decrease in average length was the result of an increase in the numbers of 114 to 137 cm lake sturgeon captured with no decrease in the number of fish in the larger size ranges.

Seven Peterson estimates from 1976 to 1982 ranged from 12 000 to 36 500 lake sturgeon over 114 cm with a mean of 25 300 (Table 5). These estimates were significantly higher ($p<0.05$) than five Peterson estimates from 1955 to 1959 which had a mean of 10 200 (SE = 714).

The number of lake sturgeon marked prior to the 1982 season (502) combined with the high harvest (2238) resulted in 32 tag returns. A density estimate by length interval was made to reduce bias due to the size difference of the fish marked in the netting operation and those harvested by spear. The 1982 estimate by length interval totalled 33 220 compared to 34 100 when all lengths were combined. Therefore, estimates from 1976 to 1982 may have overestimated by 2.6%.

Mean annual exploitation by spear from 1976 to 1983 was 2.5% ranging from 0.2% in 1983 to 6.4% in 1982. None of the 39 lake sturgeon harvested in

Table 4. Total number, mean length and length frequency of lake sturgeon over 114 cm captured by trap net and trawler, Lake Winnebago, 1975–1982.

Length range (cm)	1975 (%)	1976 (%)	1977 (%)	1978 (%)	1979 (%)	1980 (%)	1981 (%)	1982 (%)	Total (%)
114–125	33	33	28	31	38	53	49	43	41
127–137	24	28	32	28	24	23	25	30	27
140–150	23	21.5	17	18	17.5	10	13	16	16
152–163	11	8	12.5	13	12.5	7	8	8	9
165–175	7	8	6	8	6	5	4	3	5
Over 177	2	2.5	3.5	2	3	2	1.5	1	2
Total number	148	228	264	228	358	358	502	739	2826
Mean length	136.9	136.7	138.4	135.1	135.6	130.6	131.1	131.6	133.6

1983 had tags from the previous marking effort; however 39 fish removed from a stock of 25 300 is 0.2%. The 2.5% compares to 4.3% exploitation from 1955 to 1959, suggesting a decrease of 1.8% in the exploitation rate by spear.

Non-return of tags on marked lake sturgeon speared from Lake Winnebago was felt to be minimal during the 1976 to 1983 seasons. Spearers were made aware of the tags through news releases and personnel at the registration stations were required to collect tags as fish were registered. A ten dollar incentive was offered to spearers for tag returns in 1982 and 1983 along with spot checks of the registration stations by department personnel. These efforts did not result in a higher ratio of tag returns, suggesting most tags were recorded.

Lake sturgeon spawn in the Fox and Wolf Rivers from mid April to early May. From 1975 to 1983, spawning activity occurred over 4 to 10 day periods with 2 to 3 days of peak activity when the majority of lake sturgeon spawn (Table 6). Spawning was initiated in the Wolf River system when water temperatures of 11.7° C were reached and maintained. However, from 1975 to 1983 spawning activity at Fox River sites occurred when a water temperature of 14 to 16° C was reached, probably due to a rapid increase in temperatures (Priegel & Wirth 1974). If water temperatures drop below 11.7° C, spawning activity may be segmented. In 1979 temperatures dropped, spawning halted on 26 April and did not resume until 8 May.

From 1975 to 1983, 3380 lake sturgeon were marked during the spawning run. The number captured per year averaged 376, ranging from 106 in 1975 to 577 in 1979. Of the total, 2925 mature lake sturgeon were marked at 12 spawning sites in the Wolf River system and 455 at two sites on the Fox River. Netting operations identified three major sites on the Wolf River and major site on the Fox River (Fig. 1). Numerous minor spawning were identified through netting and observation on the Wolf and Embarrass rivers.

Over the 9 year period, 436 females and 2467 males (1 to 5.7 ratio) were identified at the Wolf River sites. The mean length of females was 160.3 cm, ranging from 129.5 to 200.7 cm. Males had a mean length of 133.4 cm, ranging from 82.6 cm to 175.3 cm. The mean lengths were consistent over this period (Table 7). Combined lengths of male and female lake sturgeon give an insight to the length frequency of the adult stock. Only 4.5% of the lake sturgeon captured at spawning sites were less than 114 cm, 20.5% ranged 114 to 125 cm, 32.5% ranged 127 to 137 cm, 23.6% ranged 140 to 150 cm, 12.4% ranged 152 to 163 cm, 4.9% ranged 165 to 175 cm and 1.8% were over 177 cm.

Netting efforts from 1955 to 1959 captured 675 lake sturgeon at seven spawning sites in the Winnebago system. Of this total, 96 females were identified which had a mean length of 163.6 cm (SE = 1.2) and 550 males which had a mean length of 136.9 cm (SE = 0.5). A comparison of means indi-

Table 5. Density estimate of lake sturgeon over 114 cm in Lake Winnebago, 1976–1983.

Season	Total harvest	Fish marked previous year	Tag returns (%)	P.E.	95% conf. limits
1976	936	148	6 (4.1)	19800	± 70
1977	287	228	2 (0.9)	21900	± 99
1978	1246	264	10 (3.8)	29900	± 57
1979	421	228	7 (3.1)	12000	± 66
1980	763	359	11 (3.1)	22800	± 55
1981	407	358	3 (0.8)	36500	± 89
1982*	2238	502	32 (6.4)	34100	± 34
1983*	39	739	0 (0)	–	–
	6337	2826	71 (2.5)	25300	

* $10.00 incentive system for tag returns

Table 6. Spawning dates of lake sturgeon in Fox and Wolf Rivers, 1975–1983.

Year	Spawning dates	Length	Peak spawning	
			Fox River	Wolf River
1975	May 1–6	6 days	May 1	May 5
1976	April 14–19, May 3	7 days	April 15	April 17–18
1977	Aril 12–20	9 days	April 13	April 18–19
1978	April 26–30, May 1	6 days	April 28	April 29–30
1979	April 24–26, May 8–11	7 days	not documented	April 25, May 9–10
1980	April 21–24, May 1–3	6 days	April 21	April 22
1981	April 10–16, May 5–7	10 days	April 11	April 14–15
1982	April 26, May 1	6 days	April 26	April 27–28
1983	April 26–29	4 days	April 26	April 26–28

cated a significantly lower mean length in males from the 1975 to 1983 period compared to males from the 1955 to 1959 period ($p<0.05$). However, there was not a significant difference in the mean length of females from the two periods.

Of the 3380 lake sturgeon captured at spawning sites from 1975 to 1983, 260 had been marked in previous years. One hundred eighty-seven had been marked at spawning sites and 73 during other surveys (69 in Lake Winnebago, 3 in Lake Poygan and 1 in the Wolf River). Of the 187 lake sturgeon previously marked at spawning sites, 6 were females and 181 were males. Three females were recaptured after three years, two after four years and one after six years. Of 144 male recaptures since 1980, 11% returned in one year, 34% in two years, 17% in three years, 21% in four years and 17% from five to seven years later. One male was recaptured at the same spawning site three consecutive years. Four returns were from tagging efforts from 1956 to 1961.

Tag returns indicate that lake sturgeon tend to return to the same spawning areas, as 162 of 187 (87%) marked fish returned to the same site or within a few kilometers if rock riprap had been recently placed on a nearby river bend. The major spawning site below the Eureka Dam on the Fox River was created in 1977 when rock riprap was placed along a 400 foot bank to stop erosion. Prior to 1977, 89 lake sturgeon had been marked at a site 0.8 km downstream. From 1977 to 1983, 10 of the 89 fish were captured at the new site. Concentrations of spawning lake sturgeon were not observed at the old site since 1976.

Table 7. Number, mean length (cm), and sex ratio of lake sturgeon, Wolf River, 1975–1983.

Year	Females			Males			Sex ratio
	Number	Mean length	SE	Number	Mean length	SE	
1975	9	168.2	3.4	61	133.9	1.8	6.8:1
1976	37	159.3	2.2	272	133.1	0.8	7.4:1
1977	36	157.7	1.7	245	130.3	0.8	6.8:1
1978	59	161.8	1.6	268	132.1	0.8	4.5:1
1979	88	159.8	1.5	487	133.1	0.6	5.5:1
1980	35	159.5	2.6	261	136.4	0.8	7.5:1
1981	55	159.3	1.9	319	133.6	0.7	5.8:1
1982	58	162.1	1.5	222	132.8	0.9	3.8:1
1983	59	160.3	1.5	332	135.1	0.7	5.6:1
Total	436	160.3	0.7	2467	133.4	0.3	5.7:1

Concentrations of spawning lake sturgeon were documented at several sites on the Wolf and Embarrass rivers after rock riprap had been placed on outside river bends. These new sites were attracting small numbers of lake sturgeon which normally spawned at other sites in the system.

Discussion

Data on the annual harvest, length, weight, age and tag returns collected through the mandatory registration system from 1955 to 1983 on lake sturgeon harvested from Lake Winnebago gives no indication of a decline in the stock over 29 seasons. The extreme fluctuation in the annual harvest made it necessary to look for trends over a period of seasons in order to determine if change had occurred in the population of this long lived species.

The expected result of the regulation change in 1974 which increased the minimum length of harvestable size lake sturgeon was an 11% decrease in the harvest. This regulation was instituted to bring the harvest in line with a conservative estimate of annual recruitment (540 fish) based upon density estimates from 1955 to 1959. However, the mean annual harvest from 1974 to 1983 was 698, a 19.5% increase. During this period the mean number of spearers was 5297, a 32.7% increase over the previous 14 years. Spearer success remained the same (13.2%) for both periods. Therefore, the 1304 additional spearers from 1974 to 1983 were responsible for a mean annual harvest of 172 fish. If spearer numbers had remained at the levels present prior to the regulation change, then the mean annual harvest would have been 526 fish, in line with projections. It appears the impact of the increased length limit was offset by the 32.7% increase in spearing pressure. Future regulation changes should consider trends in spearer numbers. If a regulation change becomes necessary to reduce the harvest, a shortened season may be more effective than an increased length limit.

Although spearer success in a single season can be influenced by environmental factors, the mean success over a period of seasons will likely reflect changes in the lake sturgeon population. A drastic decline in the stock should result in a decrease in spearer success, especially during seasons with favorable spearing conditions. Mean success of spearers on Lake Winnebago was the same for the 1960 to 1973 and 1974 to 1983 periods, even though spearing pressure had increased and the length limit change reduced the amount of available legal size lake sturgeon. An increase in fishing pressure combined with a higher length limit should result in a decrease in spearer success unless there was an increase in the stock.

There was no systematic decline in the mean length of lake sturgeon harvested from 1955 to 1967 as reported by Priegel & Wirth (1975). From 1974 to 1983, there was no trend toward decrease in the mean length or shift in the length frequency of the harvest to indicate a decline in the number of larger, older fish. The higher percentage of fish in the 114 to 150 cm ranges was attributed to the length limit change in 1974. Over 29 seasons, the frequency of lake sturgeon over 150 cm did not decline. Mean length of the high harvest in 1982 was significantly lower than the length of the fish registered in the low harvest 1983 season. It was felt, during clear water seasons which produce high harvests, spearers were able to distinguish and spear smaller fish compared to seasons with poor clarity when the mean length was larger.

There was no significant difference in the length-weight relationship between the two periods, 1955 to 1967 and 1974 to 1983. If food supplies remain constant, and the stock is at carrying capacity then a change in density should result in changes in the length-weight relationship. However, the Lake Winnebago population did not exhibit any significant change over 29 years.

During the netting operation on Lake Winnebago, the mean length of the lake sturgeon decreased from 1975 to 1982. However, the decrease was not a result of decline in the catch of lake sturgeon over 150 cm, but a substantial increase in the number of 114 to 137 cm fish. Assuming equal effort each year, this suggests increasing numbers of small, young lake sturgeon entering the harvestable stock.

Lake sturgeon aged from the 1975, 1976 and 1981

harvests indicate a shift to younger fish compared to the 1955 to 1967 period. However, there was not a corresponding decline in the harvest, but a 19.5% increase over ten seasons from 1974 to 1983. Density estimates from 1976 to 1982 indicate 24 600 lake sturgeon over 114 cm in Lake Winnebago, a substantial increase from the 10 200 fish estimated from 1955 to 1959. Density estimates were made using the same methods for both periods. These estimates met most of the assumptions of a Peterson estimate. The small number of fish marked in a single year resulted in wide confidence limits. However, the seven estimates from 1976 to 1982 were significantly higher than the five estimates from 1955 to 1959. Some tag loss occurs but it was assumed that tag loss was similar for both periods. Effort was made to assure maximum return of tags for both periods and the amount of non-return was assumed to be similar. During the 1955 to 1959 period, the size of marked fish was considered similar to the size of harvested fish. From 1976 to 1982, the size of the fish marked was smaller than the harvest and the 1982 estimate by length interval suggested a 2.6% over estimate due to this difference. Therefore, the mean estimate of 25 300 fish from 1976 to 1982, was adjusted to 24 600 fish to allow for this bias.

Priegel & Wirth (1975) utilized two methods for determining the rate of recruitment into the harvestable stock. A simple arithmetic mean of the age representation for 15 to 19 year old lake sturgeon harvested from 1955 to 1967 gave a conservative estimate of 4.7% recruitment. An estimate of actual recruitment based upon a catch curve for age 16 to 36 fish harvested during the same period equalled 5.4%. These estimates assume recruitment was equal to the mortality rate. The same estimates of recruitment based upon the 1975, 1976 and 1981 age samples, give a mean age representation of age 15 to 19 year olds equal to 6.3% and a 10.5% rate based upon the catch curve.

The estimates of recruitment assume that the increase in mortality from 5.4 to 10.5% is a function of increasing recruitment. There is no evidence to suggest that the mortality rate has increased in the Lake Winnebago stock. The exploitation rate by spear harvest decreased from 4.3% (1955 to 1959) to 2.5% from 1976 to 1983. There was also no evidence of an increase in natural mortality or illegal harvest. Since the mid 1970s, there has been a substantial increase in law enforcement effort to deter illegal harvest. Data collected on the Lake Winnebago lake sturgeon stock from 1974 to 1983 indicates the number over 114 cm has increased from 10 200 in the late 1950s to 24 600. A conservative estimate of annual recruitment is 6.3%.

A vital check on the lake sturgeon of the Winnebago system is the status of the spawning stock. From 1975 to 1983, an annual mean of 376 mature lake sturgeon were captured at 14 spawning sites in the system. This was a substantial increase over a mean catch of 135 fish per year during similar netting from 1955 to 1959. While effort by handheld dip net was not documented, the higher catch rate in the 1975 to 1983 period suggests an increase in the spawning stock. Within the nine year period from 1975 to 1983, there was no trend toward decreasing catch, especially following the high spearing harvests in 1978 and 1982.

Mean length and length distribution of male and female lake sturgeon remained consistent from 1975 to 1983 giving no indication of a decline in the stock. The status of the female stock is probably more critical than the males due to the fact females mature at an older, larger size. Males mature at 14 to 16 years of age and females at 24 to 26 years as reported by Priegel & Wirth (1974). The minimum length limit of 114 cm protects most males until they reach maturity, however, most females mature at 140 cm and are susceptible to legal harvest approximately ten years prior to spawning for the first time. A comparison of mean lengths for females from the 1955 to 1959 period and 1975 to 1983 period was not significantly different. The mean number of females marked at spawning sites increased from 19 per year in the 1950s to 48 per year from 1975 to 1983, suggesting an increase in the number of females.

Mean lengths of males marked in the 1955 to 1959 and 1975 to 1983 periods, indicated a significant decline from 136.9 cm to 133.4 cm respectively. However, the mean catch per year increased from 110 in the 1950s to 274 from 1975 to 1983. The decrease in mean length is probably a function of an increase in the number of young males entering the spawning stock.

Netting of spawning lake sturgeon occurred at only seven sites in the 1950s. Many of these sites had been artifically created by rock placed on the outside bends of rivers as reported by Priegel & Wirth (1974). There are a limited number of natural rock sites suitable to lake sturgeon as spawning habitat in the system. Most of the natural sites occur in the two small tributaries of the Wolf River and were not extensively used by lake sturgeon each year. The practice of placing rock rip rap on river banks by private land owners to prevent erosion gained popularity in the 1950s and 1960s. Recent netting operations documented the creation of several new sites in the 1970s and early 1980s. Tag returns on these newly created sites suggests lake sturgeon were attracted to recently placed rock and less likely to spawn on rock which had become covered by silt, debris or heavy algal growth. Law enforcement personnel who patrol the river system during the spawning period have noted approximately 50 sites now being utilized by lake sturgeon. Many of these were minor sites not utilized by spawning fish every year.

It is apparent that the number and possibly quality of lake sturgeon spawning sites has increased since the 1950s. This appears to be the major reason for the present higher rate of recruitment into the harvestable stock of Lake Winnebago. Future studies on the lake sturgeon should attempt to evaluate the impact new and refurbished spawning sites have on the recruitment levels. It is also essential that the tributary rivers of the Winnebago system be protected against such encroachment as dams which would block spawning migrations. The artificial creation of lake sturgeon spawning sites may have application in other waters which have limited natural reproduction.

Due to the unique nature of this fish species, conservative management practices should continue on the Winnebago population. The spear fishery for lake sturgeon should be managed as a trophy sport. In light of the increased lake sturgeon stock, regulations on the Lake Winnebago spearing season should not become more liberal, especially, due to the increasing number of spearers each season. It is essential that the department continue to monitor the numbers of lake sturgeon in the Winnebago system.

Acknowledgements

We wish to acknowledge the individuals who assisted with various portions of this study. John O'Brien, Area Fish Technician assisted with most of the data collection and tabulation; Thomas Thuemler, Area Fish Manager at Marinette helped collect data and provide valuable advice on data analysis and reporting; and the Lake Michigan District operations crew at Calumet Harbor marked lake sturgeon in addition to normal netting operations on Lake Winnebago. Many other Wisconsin Department of Natural Resources personnel from various levels and disciplines, too numerous to name, provided valuable assistance with this study. Personnel at privately owned registration stations played an important role by collecting accurate data on the harvest. The ten dollar incentive for tag returns in 1982 and 1983 was offered by a group of concerned sportsmen called 'Sturgeon For Tomorrow'.

References cited

Priegel, G.R. 1971. Evaluation of intensive freshwater drum removal in Lake Winnebago, Wisconsin, 1955–1966. Wisconsin Department Natural Resources, Technical Bulletin 45. 28 pp.

Priegel, G.R. & T.L. Wirth. 1974. The lake sturgeon: Its life history, ecology and management. Wisconsin Department of Natural Resources Publ. 4–36000 (74). 20 pp.

Priegel, G.R. & T.L. Wirth. 1975. Lake sturgeon harvest, growth and recruitment in Lake Winnebago, Wisconsin. Wisconsin Department of Natural Resources Technical Bulletin 83. 25 pp.

Priegel, G.R. & T.L. Wirth. 1978. Lake sturgeon populations, growth and exploitation in Lakes Poygan, Winneconne and Butte des Morts, Wisconsin. Wisconsin Department of Natural Resources Technical Bulletin 107. 23 pp.

Probst, R.T. & E.L. Cooper. 1955. Age, growth and production of the lake sturgeon, *Acipenser fulvescens*, in the Lake Winnebago region, Wisconsin. Trans. Amer. Fish. Soc. 84: 207–227.

Ricker, W.E. 1975. Computation and interpretation of biological statistics of fish populations. Fish. Res. Board Can. Bull. 191. 382 pp.

Wirth, T. 1959. Winnebago: The big lake. Wisconsin Conservation Bulletin 24: 15–19.

Received 15.5.1984   Accepted 25.2.1985

Epilogue: a perspective on sturgeon culture

Sergei I. Doroshov & Frederick P. Binkowski

Sturgeons (Acipenseriformes, Chondrostei) live in the inland waters and coastal areas of the northern hemisphere. Two families, Acipenseridae and Polyodontidae, with 25 species, have survived to the present time.

Large brackishwater lakes, such as the Caspian, Black and Azov seas provide refuge for still abundant stocks and supply three quarters of the world sturgeon catch, ranging from 20 to 40 thousand tons a year. Stellate sturgeon, *Acipenser stellatus*, Russian sturgeon, *A. güldenstädti*, hausen, *Huso huso*, are species of major commercial importance. Schip, *A. nudiventris*, Siberian sturgeon, *A. baeri*, white sturgeon, *A. transmontanus*, Atlantic sturgeon, *A. oxyrhynchus*, lake sturgeon, *A. fulvescens*, Chinese sturgeon, *A. sinensis*, kaluga, *H. dauricus*, sterlet, *A. ruthenus* and North American paddlefish, *Polyodon spathula*, support modest fisheries. European Atlantic sturgeon, *A. sturio*, North American shortnose sturgeon, *A. brevirostrum*, one of Chinese sturgeons, *A. dabryanus* and paddlefish, *Psephurus gladius*, are endangered species, probably near extinction.

Two factors have contributed to recent sturgeon declines: overharvesting and industrialization. Low reproductive rates (late sexual maturation and high juvenile mortality), and vulnerability to fishing gears aggravated the impact of early non-regulated fisheries. River damming, eutrophication of spawning sites and water pollution have eliminated or substantially reduced natural reproduction in most geographic areas.

Ironically, these ancient fish, most vulnerable in their reproductive phase, are sought by man mainly for their eggs. Caviar (black roe) is a major commercial attraction, but boneless flesh is also highly prized. Sport fishing for sturgeon is very popular in North America.

This epilogue provides brief information on sturgeon culture and related research. Comprehensive reviews of sturgeon culture and hatchery techniques were published by Detlaf & Ginzburg (1954), Kozin (1964), Milstein (1972), and Detlaf et al. (1981).

Sturgeon culture

Fish culturists started to breed sturgeon in the hatchery 100 years ago. Wild stock enhancement through artificial propagation of wild broodfish dominated their approach to culture. Breeding of North American and Eurasian species was attempted during 1880–1920 (reviewed by Leach 1920, Kozin 1964). Success was variable. Fishculturists, unfamiliar with sturgeon reproductive biology at that time, encountered many problems. At the beginning of the century, hatchery production of sturgeons was abandoned in North America and all further progress was made in the USSR.

Industrial sturgeon culture today consists of 'ranching' of wild stocks in the Caspian and Azov seas. Hatcheries were established in the late 1930s. During 1950–1970 the rivers of the Caspian, Azov and Black seas were dammed and access to most sturgeon spawning grounds lost. Artificial propagation was adopted as a main approach to sturgeon management (Kozin 1964).

Hatchery releases reached 5 million young-of-the-year annually during 1955–1960, 21 million in 1960–1970 and 60–100 million in the last two decades (Marti 1979). Three species, hausen, stellate and Russian sturgeons, are propagated on a large scale. Schip, siberian sturgeon and sterlet are also bred in hatcheries.

Management strategy includes artificial spawning of wild-caught fish, stocking the juveniles in river deltas and harvest of sexually mature adults migrating into rivers for spawning after 10–20 year growout in the sea. Sea fishery for sturgeon is banned. Production of caviar is a target of this management scheme.

Most sturgeon species migrate into rivers and spawn in spring (although some stocks overwinter in rivers and some spawn during the fall). The hatchery cycle is of short duration, from early spring to mid-summer, in most hatcheries.

Ripe broodfish are selected from commercial seine catches and brought to the hatcheries. They are induced to spawn by injections of homogenized sturgeon pituitary glands, procured from the commercial catch. Eggs are inseminated in vitro and de-adhered by bathing in silt suspension. Inseminated eggs are then incubated in troughs resembling channel catfish incubators used in the United States.

Methods of rearing vary in different hatcheries. In some, larvae are stocked in fertilized earthen ponds and harvested as 1–3 g body weight juveniles, 30–45 days after stocking. In the others, larvae are first reared in tanks to 0.1–0.3 g body weight, using live foods cultured in the hatchery (*Enchytreus albidus, Artemia salina, Daphnia* and *Moina* species). After 15–30 days growout in tanks, young are stocked in ponds for an additional 20–30 day rearing period.

Survival from fertilized eggs to stocking size young (body weight 1 to 3 g) is low, compared with rainbow trout or channel catfish, perhaps 20 to 30% in industrial scale culture. High mortality often occurs during the egg incubation and rearing of larvae. However, fecundity of the broodfish is high, ranging from 200 thousand eggs in stellate sturgeon to a million in hausen. Hence, numerous progeny can be obtained from single successfully spawned female.

Young are stocked in river deltas during the summer. There are no effective marking techniques to determine commercial return of fish stocked. Their survival is monitored by periodic area-density observations conducted in the sea. Survival to adults returning for spawning is projected to be 1 to 3%. During the late 1970s the annual sturgeon catch in the Caspian Sea basin was 26 thousand metric tons, yielding 1750 tons of caviar (Marti 1979).

Commercial growout in ponds and cages was attempted with the small freshwater species, sterlet, but has never emerged as a practical farming technique. A fertile intergeneric hybrid (hausen × sterlet, called bester) was raised and bred for two generations in ponds of southern Russia (Burtzev & Serebryakova 1980). The first generation exhibited good growth and adaptability to culture conditions, but growth and reproductive performance of the F2 was inferior. Commercial growout of bester in ponds and cages was recently established. Three to four thousand tons of fish, 0.8–1.2 kg body weight, are produced annually. Harvest is modest, 15 to 20 kg per square meter of cage and less than one ton per hectare of pond (Romanycheva & Salnikov 1979).

Domestication was recently started in France and the United States. Siberian sturgeon, imported from the USSR, were raised in France to age 7 years. Ripe males and a few ripe females were stripped, using common carp pituitary extracts and LH-RH analogue for the induction of gamete maturation (Patrick Williot, personal communication).

Wild lake sturgeon have been stripped in the laboratory and F1 progeny raised to market size in cages (University of Wisconsin). Paddlefish young produced in the hatchery are raised in warmwater ponds of southeast United States, in polyculture with other species (University of Auburn). Hatchery breeding of the Atlantic and shortnose sturgeon has been initiated in South Carolina (Smith, this volume).

At the University of California, Davis, and in some private fish farms white sturgeon has been raised in tanks to an average body weight of 0.8–1.5 kg within 18 months after hatching. Harvest was 50–60 kg per cubic meter and food conversion on dry trout diet was approximately 2:1. The University of California maintains several year-classes of captive white sturgeon for breeding. All males matured at 4–5 years of age (8 to 16 kg body weight) and were crossed with wild females to produce a 'semi-domestic' progeny. However, no ripe females were observed to age 5 years in captivity.

Growth and food conversion of domestically raised sturgeon are satisfactory for commercial culture but their yield at harvest is not as high as other cultured fish species, and their successful breeding in captivity is still uncertain. Yield can be improved through breeding and selection for growth, body conformation and tolerance to high density. Sturgeons are iteroparous and their broodstocks can be used for repeated spawning over a long period of time. Although the generation intervals are long, the efficiencies of breeding programs can be achieved through mating of different age groups and evaluation of parents by the performance of their progenies. Males, producing large volumes of semen, can be kept in minimal number and their semen cryopreserved (Burtzev & Serebryakova 1969).

Research applied to culture techniques

Egg adhesivity and difficulties in obtaining naturally ovulating females were two major problems in the early days of sturgeon culture. Inseminated sturgeon eggs secrete an adhesive glycoprotein coat (Cherr & Clark, this volume). When incubated 'en masse', they adhere to each other and die from suffocation and fungal growth.

The first problem was promptly solved by application of the simple technique of mudding, i.e. coating inseminated sturgeon eggs with natural silt particles. The second was more difficult to solve. Gerbilsky (1941) and his co-workers developed a hormonal spawning induction technique, using sturgeon hypophyses. This new technique was implemented with caution (Sadov 1950) but was proven highly successful and is now used in all sturgeon hatcheries.

Further research on sturgeon gametogenesis, seasonality of reproduction and spawning migrations of different ecotypes provided background for various spawning induction schemes (Detlaf et al. 1981). However, the functional side of reproductive endocrinology of sturgeon has never been investigated in depth. Relatively recently some progress was made through the identification of sturgeon gonadotropins and in vitro assays of various reproductive hormones (Bursawa-Gérard et al. 1975, Lutes, this volume). Sturgeon hypophyses are still used for spawning in all Soviet hatcheries. Hypophyses of common carp and synthetic LH-RH analogues are used in the United States, China and France.

T.A. Detlaf, A.S. Ginzburg and their co-workers (Detlaf & Ginzburg 1954, Ginzburg 1968, Detlaf et al. 1981) have provided major contributions to sturgeon culture, and insights into many aspects of reproductive biology, particularly the oocyte maturation, gamete physiology and embryogenesis. Sturgeon spermatozoa possess an acrosome and exhibit an acrosome reaction during fertilization. Ova have multiple micropyles and polyspermic fertilization is possible. Egg insemination by diluted semen helps to overcome the polyspermic effect and to achieve better production of larvae in the hatcheries.

Cleavage of sturgeon eggs is holoblastic and all patterns of early development differ from the meroblastic cleavage of teleosts, resembling more closely those of anuran Amphibia (Ballard & Ginzburg 1980). Yolk is intracellular and kept in crystalline form throughout embryogenesis. Embryonic development is fast and occurs in relatively warm water, 5 to 10 days at 10 to 20° C, for different species (Wang et al., this volume). Major mortalities and malformations occur during morphogenetic movements of embryonic layers, i.e. during a period from differentiation of the dorsal lip to germ ring closure (Detlaf et al. 1981). Hatching is realized by the action of hatching enzyme secreted by embryonic gland cells (Zotin 1953).

Organogenesis and morphology of juvenile

sturgeon are described in numerous Russian papers, reviewed by Detlaf et al. (1981). The differentiation and function of the gastrointestinal tract is of particular interest. Endodermal cells, rich with yolk, form the gut primordium containing a large mass of cells loose in the anterior portion of the gut lumen. Yolk platelets are digested by intracellular and lumenal proteolytic enzymes. Differentiation of the gut proceeds from the posterior (spiral intestine) to the anterior (gastric) region (Schmalgauzen 1968).

Exogenous feeding starts when the stomach is differentiated and void of lumenal yolk, at 8–14 days after hatching. By this time gastric glands are functional and proteolysis is supplemented by gastric secretion of acid and pepsin (Buddington & Christofferson, this volume). During the endogenous feeding white sturgeon embryos lose 40% of the protein, 54% of the fat and about 50% of the calories of their bodies (Wang 1984). Initiation of the exogenous feeding at the proper time and on a proper diet is crucial for their survival.

Although we know that sturgeon are carnivorous, their specific nutrient requirements are not known. Prepared moist diets, based on fish meal, blood meal, silkworm pupae, with inorganic phosphorus added, are used in Soviet hatcheries with relatively poor success. White sturgeon juveniles are raised in California on salmonid moist diets with 50% protein content (Biodiet, Bioproduct, Inc.) and dry trout feed (Silvercup, Utah). Survival from the larval to juvenile period is acceptable (40 to 60% in commercial culture) but growth of fish during the first month of feeding is clearly inferior to those raised on live food (Buddington & Doroshov 1984). Research on sturgeon nutrition and formulation of sturgeon-tailored artificial diets are critical for the improvement of sturgeon culture.

Understanding feeding and learning behavior of sturgeons is equally important. White sturgeon larvae refuse to eat any artificial diets if they have been fed live food during the initiation of exogenous feeding. They behave as if 'imprinted' on live food smell and taste (Buddington & Christofferson, this volume). However, if artificial diets are offered from initiation of feeding, larvae will accept these diets and 'imprinting' will not occur, i.e. a shift to live food is possible at any later time. Apparently, they rapidly learn to recognize a most preferred food and reject all others that are very different. This learning may involve physical and chemical properties of diets, their appearance, taste and smell, as well as the post-ingestional experience of fish (Lindberg & Doroshov 1985).

Many other aspects of sturgeon culture are now just emerging and so their discussion must be speculative. For example, understanding ontogenetic changes in osmoregulatory function of sturgeons (McEnroe & Cech, this volume) is important for possible release of hatchery produced fish in tidal estuaries of North America, where salinity is variable. Knowledge of population genetics is needed for the strategies and management of hatchery stocking (Bartley et al., this volume). Research background in animal husbandry disciplines (breeding, nutrition, physiology, pathology) is inadequate or totally absent.

Conclusions

A century old sturgeon culture has resulted in an important research background on the biology of acipenserid fish and industrial scale artificial propagation, established in the basins of Caspian and Black seas. However, this 'ranching' approach has not achieved a longterm success. All components of this complex system (man and his activity, response of cultured species, changing environment) are not understood well enough to support a balanced strategy of hatchery production and stock harvest.

Although hatchery experience with many North American species is now becoming available, no efforts have been made to initiate artificial hatchery stocking outside the USSR. Facing a gap in knowledge of life history cycles and population structure, state and federal agencies of the United States prefer a more conservative approach, i.e. strict regulation of sturgeon fishery.

New attempts to domesticate sturgeon indicate that their rearing in captivity for flesh production may be feasible. Marketing and economic feasibilities are unknown, however, and sturgeon breeding in captivity is yet to come. The historic experience of aquaculture shows that no major research efforts will be devoted to these areas until a practical culture, based on wild tamed animals, proves its success. Thus, future progress of sturgeon culture remains to be seen.

References cited

Ballard, W.W. & A.S. Ginzburg. 1980. Morphogenetic movements of acipenserid embryos. J. Exp. Zool. 213: 69–103.

Buddington, R.K. & S.I. Doroshov. 1984. Feeding trials with hatchery produced white sturgeon juveniles (*Acipenser transmontanus*). Aquaculture 36: 237–243.

Burtzev, I.A. & E.V. Serebryakova. 1969. First experiments on cryopreservation of sturgeon semen. pp. 94–99. *In*: Trudy Molodykh Uchenykh, Vol. 1, VNIRO, Moscow. (In Russian).

Burtzev, I.A. & E.V. Serebryakova. 1980. Evaluation of bester broodstock (hybrid hausen, *Huso huso*, and sterlet, *Acipenser ruthenus*) by cytological characters and viability of progenies. pp. 63–70. *In*: Variability of Karyotypes, Mutagenesis and Gynogenesis in Fishes, Institute of Cytology, Akad. Nauk SSSR, Leningrad. (In Russian).

Burzawa-Gérard, E., B.F. Goncharov & Y.A. Fontaine. 1975. L'hormone gonadotrope hypophysaire d'un poisson chondrostéen, l'Esturgeon (*Acipenser stellatus* Pall.). I. Purification. Gen. Comp. Endorcinol. 27: 289–295.

Detlaf, T.A. & A.S. Ginzburg. 1954. Embryonic development of acipenserid fish (stellate, Russian sturgeon and hausen), in connection with artificial propagation. Publ. House Akad. Nauk SSSR, Moscow. 216 pp. (In Russian).

Detlaf, T.A., A.S. Ginzburg & O.I. Schmalgauzen. 1981. Development of acipenserid fish. Publ. House 'Nauka', Moscow. 224 pp. (In Russian).

Gerbilsky, N.L. 1941. Method of hypophyseal injections and its significance in fish culture. pp 5–35. *In*: Methods of Hypophyseal Injections and Its Significance in Fish Stock Enhancement Through Artificial Propagation, Publ. House Leningr. Univ., Leningrad. (In Russian).

Ginzburg, A.S. 1968. Fertilization of fish and problem of polyspermy. Publ. House 'Nauka', Moscow. 358 pp. (In Russian).

Kozin, N.I. 1964. Sturgeon of the USSR and their artificial propagation. Trudy Vses. Nauchno-issled. Inst. Morsk. Rybn. Khoz. Okeanogr. 52: 21–57. (In Russian).

Leach, G.C. 1920. Artificial propagation of sturgeon, review of sturgeon culture in the United States. Rep. U.S. Fish Comm. 1919: 3–5.

Lindberg, J.C. & S.I. Doroshov. 1985. Effects of diet change (natural and prepared) on growth and survival of white sturgeon juveniles. Abstracts of 9th Annual Larval Fish Conference, February 24–28, 1985, The University of Texas, Port Aransas.

Marti, Y.Y. 1979. Development of sturgeon stock management in the southern seas of the USSR. pp. 73–85. *In*: Biological Resources of Inland Waters of the USSR. Publ. House 'Nauka', Moscow. (In Russian).

Milstein, V.V. 1972. Sturgeon culture. Publ. House 'Pischevaya Prómyschlennost', Moscow. 129 pp. (In Russian).

Romanycheva, O.D. & N.E. Salnikov. 1979. Net pen mariculture and prospects for its development in the USSR's waters. Trudy Vses. Nauchno-issled. Inst. Morsk. Rybn. Khoz. Okeanogr. 137: 7–14. (In Russian).

Sadov, I.A. 1950. Effect of spawning and incubation techniques on egg mortality in Russian and stellate sturgeons. Trudy Inst. Morf. Zhivotnykh Severtsova 3: 3–18. (In Russian).

Schmalgauzen, O.I. 1968. Development of digestive system of acipenserid fish. pp. 40–70. *In*: Morpho-ecological Research on the Development of Fishes, Publ. House 'Nauka', Moscow. (In Russian).

Wang, Y.L. 1984. The effect of temperature on early development of white sturgeon and lake sturgeon. M.S. Thesis, University of California, Davis. 54 pp.

Zotin, A.I. 1953. Hatching enzyme of acipenserid embryos. Dokl. Akad. Nauk SSSR. 92: 685–687. (In Russian).

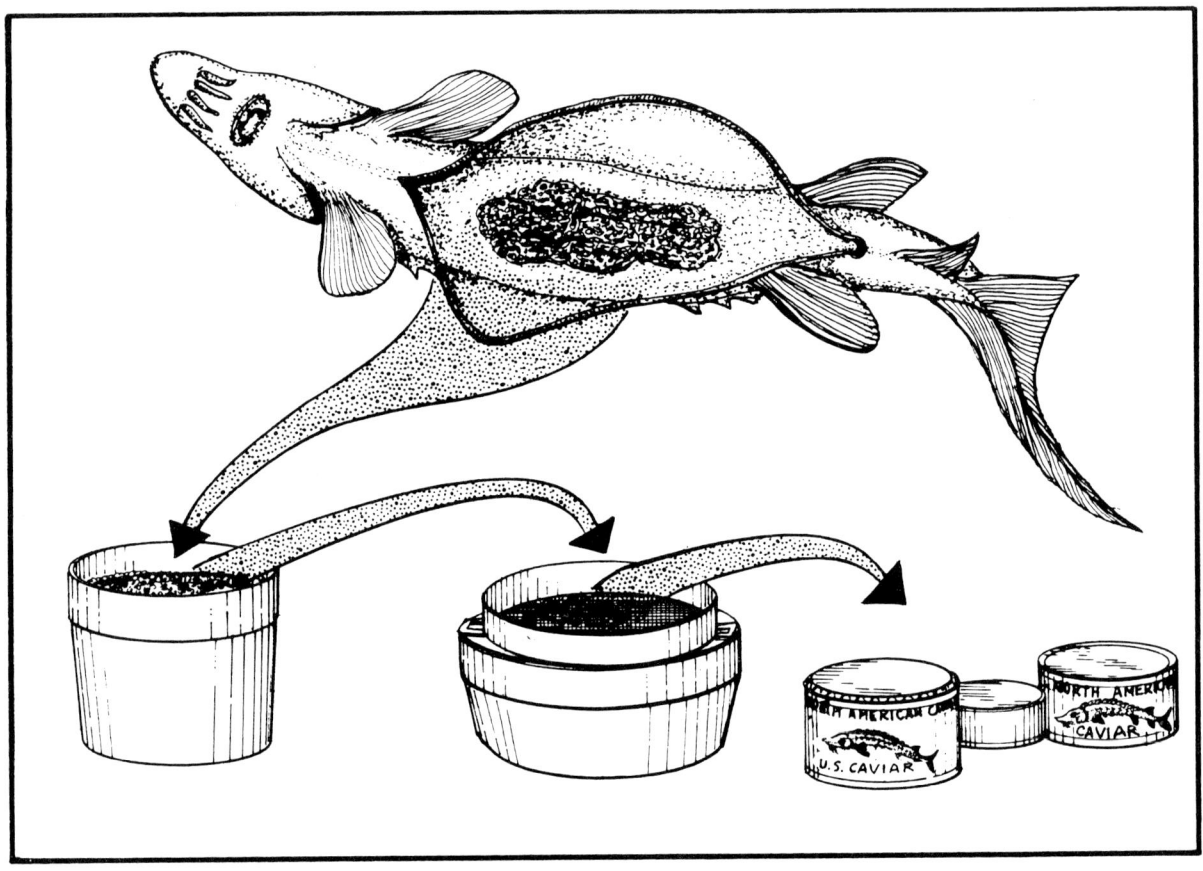

The products of sturgeon aquaculture include: high quality flesh and sturgeon roe which comprise about 20% of the body weight. This raw material is processed as caviar and sold as a gourmet food product. Sturgeon swimbladder supplies a unique binding substance used in commercial art glue.

Species and subject index

Abundance 51, 56, 61, 127–129, 131
A.C. boomshocker electrofisher units 74
Acclimation 26–29
Achlya 68
Acipenser baeri 28, 49, 116, 147
Acipenser brevirostrum 6, 8, 62, 111–117, 147
Acipenser dabryanus 147
Acipenser fulvescens:
   artificial spawning, rearing 79–85
   cross-fertilisation 11, 18
   distribution 6
   effect of temperature 43–50
   fisheries 147
   Lake Winnebago 135–146
   Menominee River 73–78
   osmoregulation 28
   spawning 111
   symposium 8
Acipenser guldenstaedti 28, 49, 147
Acipenser medirostris 6, 28, 119
Acipenser moliventris 28
Acipenser nudiventris 147
Acipenser oxyrhynchus 6, 8, 61–72, 97, 111, 147
Acipenser oxyrhynchus desotoi 62, 97, 103
Acipenser oxyrhynchus oxyrhynchus 61, 97–102
Acipenser ruthenus 28, 49, 147
Acipenser sinensis 147
Acipenser stellatus 49, 87–90, 147
Acipenser sturio 28, 147
Acipenser transmontanus:
   animo acid levels 93–95
   Columbia River 119–125
   distribution 6
   effect of temperature 43–50
   fisheries 147
   gamete interaction 11–22
   genetic structure 105–109
   Idaho 127–133
   oocyte maturation 87–92
   osmoregulation 23–30
   spawning 111
   symposium 8
Acipenseridae 28, 52, 61, 147
Acipenserids 12, 14, 21, 31–39, 43, 50, 150
Acipenseriformes 147
Acrosomal vesicle 15, 20
Acrosome reaction 11, 15–21, 149
Acrosomes 11–12, 15–18, 21
Actinopterygian fish 23
Adaptive radiation 49
Adhesiveness 12–13, 65, 80, 84, 115, 149
Age:
   determination 73–75, 127–133, 144–145
   fisheries 119–124
   length-age 76, 140–141
   long-lived 67
   sexual maturity 23, 149
   spawning 23, 122, 132, 135
   weight-age 83–84, 140–141
Alabama 70–71
Alimentary canal 32–39
Allelic variation 105–109
Allometric growth 100
Allopatric 49, 97–98, 102
Allozyme 105–108
Alveoli 21
Ambloplites rupestris 74
Amino acid levels 93–95
Amylase 36, 39
Anadromous:
   categories 27
   elevator 112
   gametes 12
   morphometrics 97
   osmoregulation 28–29
   ranching 10

spawning 23, 62, 64
susceptible 61
Anatomy 31–35, 101–103
Ancient animal 7, 9, 147
Androscoggin River 64
Animal pole 13–15, 44
Annuli 53, 74, 131
Anuran Amphibia 149
Apalachicola River 65–66, 97–98
Apical organelle 15
Aquaculture 7, 10, 31, 62, 105, 151–152
Aquarium fish 10
Art glue 63, 152
Artemia sp. 38, 81
Artemia salina 148
Artificial culture 7
Artificial propagation:
  early research 7–8
  importance 68
  methods 79
  perspective 147–150
  stock rehabilitation 69
  temperature 43
Artificial spawning:
  first success 68
  hatchery 79
  injections 148
  oogenesis 87–91
  perspective 149
Asian species 9
Asparagine 93–94
Atlantic Ocean 9, 61–72, 97, 101–102
Atlantic sturgeon:
  breeding 148
  distribution 6
  fisheries 147
  management 61–72
  migration 9
  morphometrics 97
  spawning 111
  symposium 8
  temperature 43
Azov Sea 43, 147

Bag limit 77, 120
Bay of Fundy 62
Behavior 7, 10, 111, 115–116, 122, 150
Beluga 49
Benthic 31, 37, 65–66
Benthophagy 31, 37
Bermuda 62
Bester 148
Binding proteins 15
Biochemistry 12
Biological characteristics 70
Biological data 7, 62, 70, 150

Biological studies 69
Biology 7–8, 51, 61, 64–68, 150
'Bioscrews' 14–15
Black roe 147
Black Sea 147–148, 150
Blood electrolytes 23
Blood samples 23–27, 93
Body proportions 99–101
Bonneville Dam 119–125
Breeding 147–151
Broadcast spawning 12, 17–18, 21, 65
British Columbia 23
Broodstocks 149
Brooding/brooders 12, 21, 123, 147–148

Caesarian section 79–80, 84
California 8, 10, 23, 150
Calories lost 150
Canada 7, 10, 62–63, 68–71, 97
Cannulation 23–25
Carassius carassius 90
Carbohydrase activity 31, 36
Carbohydrates 11, 13, 18, 36, 39
Carnivores 31, 36, 39, 150
Caspian Sea 43, 147–148, 150
Catabolism 93, 95
Catacholamine 87–88
Catch and release 127–128, 132–133
Catheterization 86
Catostomus commersoni 74
Caudal section 26
Caviar 7, 63, 122, 147–148, 152
Census 74, 124, 128
Chapman's modifications 124
Chemical communication 87
Chesapeake Bay 62, 65
China 8, 10, 149
Chinese sturgeon 147
Chi-square analysis 68
Chloride levels 28
Chondrostean(tei):
  Acipenseriformes 147
  amino acids 93
  digestion 31
  evolution 89
  feeding 31
  fertilization 12
  morphometrics 97
  oocytes 87
  osmoregulation 23–24, 27
  spawning 111
  temperature 43
Chorion 11–12, 14
Chymotrypsin-like 13
Cleavage 43, 45, 50, 65, 149
Closed areas 69, 133

Closed seasons 69
Clupea pallasi 93
Coastal environment 61–66
Coelomic fluid 12, 16
Columbia River 9–10, 105–106, 119–125
Commercial value:
    bester 148
    caviar/roe 7, 63, 122, 147–148, 152
    demand 9, 152
    feasibility 10, 150–152
    flesh 7, 63, 119, 147, 150, 152
    gourmet food 10, 122, 147, 152
    harvest 61, 105, 119–123, 147–148
    production 10, 147–148
    products 63, 152
    species 147
Connecticut 70–71
Connecticut River 64, 111–117
Conservation 9, 70, 146, 150
Consortium 10
Constructions permits 62
Controlled culture systems 62
Cortical aveoli 20–21
Cortical reaction 20–21
Corticosteroids 87, 89
Cortisol 28–29
Creel census 73–77, 124
Critical habitats 68
Cross-fertilization 11, 18
Cryopreserved 149
Culture:
    Atlantic sturgeon 61–62
    background 7–10
    hatching 65
    juveniles 69
    lake sturgeon 79
    perspective 147–151
    trials 65
Cyprinids 21
Cyprinus carpio 74, 88, 93

Dams:
    environment 62, 119, 124, 127
    population decline 43, 64, 109, 147
    recruitment 9, 132–133, 146
    spawning:
        Connecticut River 111–112
        Fox River 143
        Lake Winnebago 138
        Menominee River 73–77
Daphnia sp. 79, 81, 83, 84, 148
Data base 7, 124
Delaware 62–65, 70–71
Demand 9
Density estimates:
    formula 129
    lake sturgeon:
        Lake Winnebago 135–136, 141–145
        Menominee River 75–79
    release survival 148
    white sturgeon 128–129
Density tolerance 149
Deoxygenated blood 25
Depletion 7, 9, 61, 64, 79, 119, 123
Development 79, 149–150
Developmental rates 43–50
Development stages 44–47
Diadromous 28
Diet 31, 51, 56–57, 66, 79–84, 150
Digestive system:
    anatomy 32–35, 39
    development 35, 39, 96, 150
    enzymes 31, 35–39
    physiology 31–37
    structure 31–35, 39
Dip nets 52, 80, 138, 145
Disease prevention 85
Distance 66, 123
Distribution 51–56, 127
Domestication 148–150
Dorsal scutes 47, 97–102, 106, 113
Dorsal walls 34
Dorso-ventral axis 13
Dry feed 81–84

Early development 35, 79, 96
Echinoderms 11
Ecological relationships 58
Ecology 7, 61, 97, 111
Economic difficulties 9
Ecotypes 149
Ectothermic 89
Egg(s):
    adhesiveness 12–13, 65 80, 84, 115, 149
    caviar 7, 63, 122, 147–148, 152
    collection 44, 79–80, 84
    discharge 65, 115
    envelope 11–19
    handling 79–80
    incubation 44, 48, 78–87, 148
    jelly coats 11–15
    PCB effect 124
    resorption 124
    roe 63, 147, 152
    size and color 12, 45, 65, 82
    structure 11–15
Eggwater 11, 17, 18
Electric generating stations 62, 127
Electrical stimuli 39
Electrofisher 98, 128
Electrofishing surveys 78
Electrolyte determinations 25–28

Electrophoresis 53, 56, 105, 106, 109
Electrophoretic analysis 18
Electroreception 31, 37, 39
Embryogenesis 50, 149
Embryonic 35, 48–50
Embryonic survival 43–50
Embryoic development 13, 44–50, 149
Embryos 43–50, 65, 79–84, 96, 115, 150
Enchytreus albidus 148
Endangered 9, 51, 62, 69, 70, 147
Edocrine control 89
Endocrinology 149
Endodermal cells 150
Endogeneous 116, 150
Enfields Rapids Dam 64
Envelope 11–19
Enviroment:
   culture 150
   gamete interaction 11
   hybridization 51, 58
   morphology 109
   osmoregulation 24
   recruitment 133
   reproduction 116
   spearfishing 144
   temperature 49
Environmental impact statements 62
Enzyme 13–18, 37, 106–108, 149–150
   Enzyme complement 31, 35–39
Epinephrine 87–91
Erytrocytes 93
Escapement 123–124
Esophagus 32
Esox lucius 74, 90
Establish stock abundance 61, 69
Estuarine environment:
   embryos 60
   green sturgeon 119
   growth rate 124
   juveniles 66
   life cycle 61, 64
   migration 23
   osmoregulation 150
Eurasian acipenserids 43, 49
Eurasian species 147
European Atlantic sturgeon 147
European species 9
Euryhaline 28–29
Eutropication 147
Evolution 7, 16, 23, 27, 89, 147
Exocytosis 11, 15, 20–21
Exogenous 38, 47, 90, 116, 150
Exploitation:
   lake sturgeon 73, 77
   moratorium 70
   regulated 135

   spearfishing 141–145
   white sturgeon 119
Exponential equations 43, 47–48
Exponential relationship 43, 47–49
External feeding 96
Extinction 147

Fat loss 150
Fecundity 65, 148
Feeding:
   adults 56, 66
   amino acids 93
   availability 122
   Chondrosteans 31–32
   fingerlings 81
   juveniles 65–66
   larvae 81
   nutrition 10, 150
   rations 38–40
   sensory 37–38
Fertilization 11–21, 79–80, 86, 148
Filament formation 11, 16
Fingerlings 81, 96
Fish eggs 12–15
'Fish life areas' 70
Fish wheels 120
Fisheries:
   Atlantic sturgeon 61–69
   Columbia River 119–123
   lake sturgeon 73, 136
   Lake Winnebago 136
   management 7, 109, 148, 150
   Menominee River 73–77
   species 147
   white sturgeon 105, 119, 131
Flagellum 12, 16
Flesh 7, 63, 119, 147, 150
Florida 62–67, 70–71, 97
Flow regimens 58, 127, 133
Follicle 87–90
Follicular envelope 88
Follicular maturation 89
Food:
   conversion 149
   deprivation 93–95
   location 31–32, 37–39, 150
   river 51, 56–57
   supplies 119, 122, 124
Fork length 32, 67, 97–103, 113
Formulated feeds/rations 38–39
France 8, 10, 148–149
Fraser River 23, 106, 124
Functional morphology 12
Fundulus heteroclitus 90
Fungus 49, 80, 84, 115, 149

Gametes:
   collection 44, 68
   fusion 12
   interaction 11–21
   maturation 148
   structure 12, 149
Gametogenesis 149
Gang lines 120
Gar 68
Gastric glands 34–35, 150
Gastrointestinal tract 34, 47, 150
Gastrulation 45–49
Gear 52, 62–63, 120–121, 128, 147
Gear restrictions 69
Gene pool preservation 7
Generations 9, 149
Genetics:
   integrity 105
   stocking 10, 150
   structure 105
   swamping 58
   variation 105–109
Geographic distribution 6
Georgia 62–65, 70–71
Germinal vesicle breakdown (GVBD) 87–90
Gill net 63, 68, 98, 105, 111, 119–121
Gizzard 32–37
Glycoproteins 11, 13, 18, 149
Glycosylated 12
Gonodal regression 116
Gonadotropins 87–89, 149
Gourmet food fish 10, 122, 147, 152
Government agencies 10
Great Lakes 7, 9, 43
Green sturgeon 6, 119
Growth:
   Atlantic sturgeon 65–69
   hatcheries 149–150
   lake sturgeon 73, 75, 83–84, 140
   river sturgeon 51, 56–57
   white sturgeon 121–123, 127–133
Growth food 31, 119
Growth rate 69, 83, 109, 119, 122–124
Gulf of Maine 64–65
Gulf of Mexico 7, 9, 62, 97–103
Gulf of Mexico sturgeon 97–103
Gulf of St. Lawrence 61
Gustation 31
Gut 29, 32–39, 47, 150

Habitat:
   diversity 28, 56, 58
   modifications 61, 68–69, 127, 133
   preservation 9, 10
   use 64, 111–113, 115, 132
Hamilton Inlet 61

Harpoons 63
Harvest:
   Atlantic sturgeon 61
   lake sturgeon 73–77, 135–146
   Lake Winnebago 135–146
   management 70, 147–150
   Menominee River 73–77
   Pacific Northwest 105, 119–123, 131
Harvesting moratorium 61, 69–70
Hatcheries:
   early development 35
   efficiency 10, 31, 147–150
   history 147
   incubation 49
   rations 31, 39
   releases 148
   Soviets 31, 147–150
   spawning 68–69, 148
   temperature 43
   Wild Rose 78–85
Hatchery releases 148
Hatching 46, 60, 115, 149
Hatching regimes 79–84
Hausen 147–148
Head length 62
Head length–fork length 99–102
Herring roe 93–94
Heterologous fertilization 18
Heteropneustus fossilis 89
Heterozygosity 105–109
History 7, 27, 43, 51, 61–62, 68, 74, 120, 136, 147
Holoblastic 45, 50, 65, 96, 149
Holostei 12
Holotype specimens 97, 99
Holyoke Dam 111–112
Homarus americanus 109
Homologous sperm 11, 18, 21
Hook and line 73, 75, 136
Hormonal component to sea water acclimation 28–29
Hormonal spawning induction 149
Hormones 149
Hudson River 64–65
Hybridization 51–53, 58, 105
Hybrids 51–59, 148
Hydrolyzed 11
Hyperosmotic 24, 26, 28, 29
Hypertonic medium 16
Hypo-osmotic 16, 24, 26, 28
Hypothyses 149
Hypotonic medium 17
Husbandry practices 10, 150
Huso dauricus 147
Huso huso 28, 31, 49, 147

Ichthyoplankton net 122
Ictalurus punctatus 74, 132

Idaho 127–133
Imprinting 38, 85, 150
Incipient mortalities 43
Incubation periods 65, 80
Incubation temperatures 43–49, 79–84, 149
Induced spawning 87, 90
Industrialization 147
Insemination 19–20, 79–80, 86, 149
Intergeneric 148
Interior anatomy 101–103
Interspawning period 61
Interspecific fertilization 21
Intestine 32–37, 150
In vitro fertilization 18, 86, 148
In vitro incubation 87, 90
In vivo monitoring 44
Ion exchange mechanisms in gills 23
Ion excreting capacity 28
Ionic strength 16
Isoleucine 93–94
Isozymes 105, 107
Israel 8, 10
Issinglass 63
Italy 8, 10
Iteroparous species 116, 149

Japan 10
Jelly:
  coat 11–15
  hydration 12
  layer 12, 14
  release 12–13, 18
Juveniles:
  diet 25, 38, 65
  digestive system 35
  food habits 31
  fungal growth 49
  morphometrics 97
  mortality 147
  osmoregulation 23–30
  rearing 79–84
  recruitment 132–133
  stocking 69, 148–150
  survival 9

Kaluga 147
Kennebec River 64
Kootenai River 105–109, 127–133

Labrador 61, 97
Lake sturgeon:
  artificial spawning, rearing 79–85
  cross-fertilization 11, 18
  distribution 6
  domestication 148
  effect of temperature 43–50

  fisheries 147
  Lake Winnebago 9–10, 135–146
  Menominee River 73–78
  spawning 111, 116
  symposium 8
Lake Winnebago 9–10, 75, 80, 135–146
Landings 62–64, 69, 120–124
Landlocked 27–28, 105, 111, 119, 124, 127
Larvae:
  behavior 65
  food habits 31, 37–39
  rearing 79–83, 148–149
  sampling 119, 122, 124
  spawning 111
  temperature 43–50
Larval rearing 48, 79, 148
Larval stage 34–37, 48, 123, 150
Laws 69–70, 75
Lena River 49
Length:
  Atlantic sturgeon 67
  hatchery 83–84
  lake sturgeon 73–78, 135, 139–145
  subspecies differences 62, 99–101
  white sturgeon 109, 119–123, 129–133
Length-age relationship:
  Atlantic sturgeon 67
  hatchery 83
  lake sturgeon 76, 140–141
  white sturgeon 122, 130–131
Length-weight relationship:
  Atlantic sturgeon 67–68
  lake sturgeon 135–144
  white sturgeon 129–130
Leucine 93–94
LH–RH analogue 148–149
Licensing 70, 136, 138
Life cycle 39, 61, 64, 150
Life history:
  Atlantic sturgeon 61–62, 65
  carnivores 31
  lake sturgeon 135
  rations 40
  reproduction 96, 116
  research 10, 150
  river sturgeon 51
  secretory phases 36
  symposium 9
  temperature 43
  white sturgeon 119, 124, 127–128
Life span 9, 121
Limestone shoal areas 65
Lipase 36
Lipid 37, 39
Lithophilous species 116
Live food 31, 37–38, 81–85, 148, 150

Liver 33 34, 93, 95
Louisiana 70–71, 98
Low ionic strength of freshwater 16

Macrobrachium rosenbergii 109
Maine 64, 67, 69–71
Malformations 149
'Managed fishery' 69, 71
Management:
   abundance 127–132
   conservation 9, 146, 150
   depletion 7, 119, 123
   genetic structure 105, 109
   habitat 9–10, 58, 69, 115, 127, 132
   hatchery 49, 78–86, 147–150
   monitoring program 70, 119
   preservation 73
   recruitment:
     age 140
     data needed 61
     escapement 123–124
     mortality 135
     spawning 132–133
   regulations:
     Atlantic sturgeon 69–71
     conservative 150
     lake sturgeon 136–146
     range 61
     white sturgeon 119–128
   subspecies 97
Manufactured rations 31–32, 38–39
Mark-recapture method 128, 137
Maryland 62, 64, 70–71
Massachusetts 64, 70–71, 112
Materials and methods:
   amino acid levels 93–94
   distribution 52–56, 73–75, 136–138
   genetics 106
   growth, migration 121–124, 128–129
   morphometrics 98
   oocyte maturation 88
   osmoregulation 24–26
   spawning, rearing 79–81, 112–113
   temperature 43–44
Maturation inducing substance (MIS) 87–91
Maturity:
   Atlantic sturgeon 61
   captivity 149
   harvesting 148
   lake sturgeon 145
   latitude 66
   salinity tolerance 29
   sexual 23, 147–148
   susceptible 63
   white sturgeon 122, 132
Maximum catch 70, 119–120

Menominee River 73–78
Meristic comparisons 51–54, 57
Meristics 97
Meroblastic 149
Metamorphosis 35, 37, 47, 96
Michigan 10, 73–78
Microhematocrits 25
Micropterus dolomieui 74
Micropterus salmoides 74
Micropyles 11–21, 149
Microscopic anatomy 33–35
Migration 23, 28, 66, 119, 122, 127
Migration routes 68, 122
Migratory patterns 61, 64–66, 148
Milt 79–80, 84
Minimum harvest size 69–70, 120, 136
Minimum-maximum size limits:
   Columbia River 119–120
   considerations 144
   Lake Winnebago 139
   maturity 145
Mississippi 70–71
Mississippi River 51–58, 97–98, 108
Missouri 9–10
Missouri River 9, 51–58
Mitosis 50
Moina 148
Monitoring program 70
Monitoring trends 135
Monomorphic 108
Montana 127
Moratorium 61, 70
Morphogenetic movements 50, 149
Morphology 149
Morphometrics:
   characters 97–103
   comparisons 47, 51–57, 80, 97–103
   variation 97
Mortality:
   factors 68–69
   hatching 35, 81–83, 148–149
   juvenile 147
   lake sturgeon 135–136, 140, 145
   rations 38–39
   salinity 26–27
   sublegals 121
   temperature 43, 47–49
   white sturgeon 131–132
Motility 11, 16–18, 21
Mt. St. Helens 122
Movements 111, 122
Moxostoma spp. 74
Mudding eggs 149
Mugil cephalus 29

Natural habitat 61, 64

Natural harvest 7
Nesting fishes 18
Net pens 62
New Brunswick 62, 67, 111
New Hampshire 64, 70–71
New Jersey 62, 64, 70–71
New Orleans 8
New York 62, 64, 67–71
North America 6, 62, 127, 147
North American paddlefish 147
North American shortnose sturgeon 147
North Carolina 62–66, 70–71
Northern hemisphere 7, 147
Norway 8
Nova Scotia 61
Nursery areas 61, 69
Nutrient uptake activities 35–37
Nutrition 10, 150
Nutritional requirements 31–32, 39, 150

Ocean ranching operations 62
Olfaction 31, 37
Olfactory rosettes 37
Oncorhynchus keta 88
Oncorhynchus masou rhodurus 90
Oncorhynchus spp. 121
Ontogenetic changes 150
Oocyte maturation 87–90, 116, 149
Oogenesis 87
Oolemma 12, 16, 19–21
Oregon 122
Organogenesis 43, 46–50, 149
Oryzias latipes 90
Osmolality 23, 25–28
Osmoregulation 23–30, 150
Otter trawls 136, 141
Ovaries 65, 87
Over-fishing 9, 43, 64, 68, 120, 147
Overharvesting 7, 43, 77, 147
Ovulation 87–90, 149
Ovum/ova 12, 86, 149

Pacific Northwest 105–109
Pacific Ocean 9, 23, 119
Paddlefish 147–148
Palaenoscinoids 27
Pallid sturgeon 6, 8–9, 108
Pancreas 33
Paratype specimens 97, 99
Parent-progeny relationships 69, 149
Pathology 10, 150
PCB effect 124
Pectoral fin 62
Pectoral fin length – fork length 100, 102
Pennsylvania 62, 64, 69–71
Perca flavescens 74, 90

Peripheral blood 93, 95
Perivitelline space 20
Peterson formula 74, 137, 141, 145
Petromyzon marinus 68
Phenotypic 56
Phototactic 79
Phylogenetic 11
Physiological evolution 89
Physiology 9–12, 31–39, 150
Piscivory 31
Planktivory 31
Plasma electrolytes 23–28
Plasma free amino acid levels 93–95
Platichthys stellatus 29
Political difficulties 9
Pollution 43, 64, 77, 120, 133, 147
Polyculture 148
Polymorphic 106–107
Polyodon spathula 28, 31, 108, 147
Polyodontids(dae) 31, 147
Polyspermy 12, 15, 20–21, 149
Population:
    Columbia River 123
    dam effect 109, 132
    dynamics 9–10
    genetics 150
    Lake Winnebago 135, 144
    Menominee River 73–78
    mid-1800s 7
    Pacific Northwest 105
Portal blood 95
Pound nets 63
Preservation 73
Pre-spawning 87, 111–117
Private industry 10, 152
Products 63, 152
Progestagens 87–90
Progesterone 87–90
Propagation efforts 68, 79
Prostaglandins 87–91
Protease 11, 13, 18, 31, 39
Protectionist policy 9, 70, 150
Protein 12, 36, 39, 95, 150
Proteolysis 150
Psephurus gladius 147
Pseudoscaphirhynchus 28
Pyloric apparatus 101–103
Pylorus 32, 34

Radio telemetry 97, 111–117
Ranching 10, 62, 147, 150
Rare 9, 135
Rate of development 43, 149
Rations 31–32, 39
Rearing 10, 31, 48, 62, 79–85, 148, 150
Recreational fishery 73, 121

Recrudescence 116
Recruitment:
　lake sturgeon 135, 140–146
　reduced by dams 9
　white sturgeon 119–124, 132–133
Recruitment to stocks 61, 69
Rectum 32–34
Registration 75, 77, 136–144
Regulations:
　Atlantic sturgeon 61, 69–70
　catch-release 127–128, 132
　conservative 150
　lake sturgeon 75, 77, 144
　spearfishing 10, 136, 139, 146
　white sturgeon 119–123
Rehabilitation 10
Repopulation efforts 77–78
Reporting 70
Reproduction:
　breeding 149
　captivity 10
　Connecticut River 111
　cycle 9
　gamete interaction 11–22
　low rates 147
　physiology 7
　shortnose sturgeon 115–116
　strategy 11–12, 21
　symposium 9
　temperature 43–50
　vulnerable 147
Research materials and methods:
　amino acid levels 93–94
　distribution 52–56, 73–75, 136–138
　genetics 106
　growth, migration 121–124, 128–129
　morphometrics 98
　oocyte maturation 88
　osmoregulation 24–26
　spawning, rearing 79–81, 112–113
　temperature 43–44
Resorption 124
Restocking programs 69
Restictions 120, 136
Rhode Island 70–71
River deltas 148
River discharge 111–113, 115
River habitat modifications 10, 51
Riverine environment 60–66, 71
Rock rip rap 135, 143, 146
Rod and reel 128
Roe 63, 147, 152
Russian sturgeon 49, 147–148

Sacramento River 23, 44, 49
St. John River 62–63, 97, 111

St. Lawrence River 61, 65–67
Saline areas 66
Salinity:
　blood samples 25
　body size 29
　cannulation 25
　chloride cells in gills 29
　cortisol 28–29
　hormones 28–29
　ion exchange 23
　ion excreting capacity 28
　maturational event 29
　osmolality 23, 26
　osmoregulation 23–30, 150
　plasma electrolytes 25–28
　surface to volume ratio 29
　tolerance 23–29
Salmo gairdneri 90, 93, 105, 121
Salmon gill net 105, 119
Salmonids 18, 21, 24
Salvelinus fontinalis 90
San Francisco Bay 23–25, 43, 88, 93, 106
San Pablo Bay 124
Saprolegnia 68
Scaphirhynchus albus 6, 8, 51–59, 108
Scaphirhynchus platorynchus:
　distribution 6
　electrophoretic 108
　hybridization 51–59
　spawning 111
　symposium 8
Schip 147–148
Schnabel's formula 124
Scutes 53, 62
Sea Grant Program 10
Sea lamprey 68
Sea run stock 9–10
Secretory phases 31, 36
Semen 86, 149
Semi-anadromous 27–28
Semi-domestic 149
Sensory adaptations 32, 37–38
Setline 105, 120–121, 124, 128
Set-net fishery 124
Sevrjuga 49
Sex ratio 52, 57, 68, 111, 114, 142–143
Sexual maturity 23
Shad nets 68–70
Shortnose sturgeon:
　distribution 6
　endangered 9, 147
　environments 62
　hatchery 69, 148
　spawning 111–117
　symposium 8
Shovelnose sturgeon 6, 8–9, 37, 108, 111

Sialic acid 12
Siberian sturgeon 49, 147–148
Size 29, 67, 73–76, 119, 145
Size limit 70, 73, 77, 119
Snake River 124, 127–133
South America 62
South Carolina 8, 10, 62–71, 148
South Dakota 58
Soviet scientists 8, 10, 31, 149–150
Spawning:
   areas 60, 64, 69, 111–116, 138, 142–148
   artificial 68, 79, 87–91, 148–149
   broadcast 17–18
   escapement 123–124
   frequency, periodicity 23, 63–66, 132, 143
   migrations 62–66
   population 135
   stock status 145
   techniques 68, 148
   temperature 49, 64–65, 80, 88, 122, 142
Spearing, spearfishing 10, 135–146
'Special concern' 69, 71
Species 7, 27, 61, 147
Sperm 11, 15–21, 84, 114, 149
Sperm-egg interaction 17–21
Spiral valve 33–34
Spleen 33, 97–103
Spleen length 97, 101–103
Spleen length–fork length 62, 97, 101–103
Sport 7, 9–10, 119–124, 135, 146–147
Stages of development 44, 48
Stake row nets 63
Starch gel 105–106
Status 51, 69, 98, 119, 127
Stellate sturgeon 49, 87–89, 147–148
Sterlet 49, 147–148
Stizostedion canadense 58
Stizostedion vitreum 58, 73
Stock rehabilitation 69
Stocks, stocking:
   abundant 147
   Atlantic sturgeon 61, 69
   Gulf of Mexico 97, 102
   harvest 150
   juveniles 148
   lake sturgeon 73, 78–79, 84
   management 105
   overharvested 7
   sea run 10
   white sturgeon 119, 122–123
   wild 147
Stomach 32–37, 150
Strict regulation 150
Sturgeon:
   reproduction:
      natural 11–21, 73, 147

      artificial 43–50, 68, 79–91, 148
   life history:
      perspective 148–150
      hatching 35, 65
      embryos 45–49, 82–84
      juveniles 23–29, 65–66
      maturity 29, 66, 119, 149
      adults 66–68
      habitat 9, 28, 56–58
      behavior 115–116, 150
      age/mortality 35, 67, 135, 140, 145
   temperature effects 43–50
   salinity acclimatization 23–30
   digestive system:
      anatomy 32–37, 150
      feeding 38–39, 56–57, 66, 150
   fisheries:
      historical 61, 120
      commercial 105, 121, 123, 147–148, 152
      recreational 73, 123, 135
   management:
      research, culture 147–151
      stocking, hatcheries 79–86
      laws, regulations 69–71
Sturgeon demand 9, 152
Sturgeon egg jelly 13
Sturgeon eggs 11–16, 44, 63, 79–80
Sturgeon oil 63
Subspecies 61–62, 97–103
Substrate types 58, 111–115, 132
Suisun Marsh 28
Survival:
   ancient fish 7
   dry diets 84
   embryo 43–44
   hatchery 78
   increase 35
   juvenile 79
   larval 148, 150
   salinity 26
Sustained yield 135
Swim bladdder 32, 35, 63, 152

Tagging:
   Atlantic sturgeon 66
   Gulf of Mexico 97
   lake sturgeon 74, 136–145
   shortnose sturgeon 113
   white sturgeon 119–124, 128
Tagging techniques 79, 81, 84
Taste buds 37
Taxonomy 7, 9, 97, 103
Telemetry 97, 111–117
Teleost 16, 23, 28–29, 50, 89–91, 149
Teleostean 87, 89, 91
Temperature:

adaptations 43, 49
growth 35, 43–50, 82–85, 131
incubation 44–50, 65, 80–84, 88, 149
spawning 49, 64–65, 80, 88, 122, 142
survival 43–44
timing of migrations 64–66, 111, 113
Teratogenic 84
Texas 70–71
Textural qualities of food 31, 39
Threatened 69–71
Threats to survival 51, 58, 68–69
Thyroxine 28
Timing of development 47–49
Tissue storage 95
Total fishing ban 69
Trammel nets 52, 63
Trap nets 136, 141
Traps 120, 128
Trash fish 7
Travel 66
Trawler 141
Trawls 63
Trophy sport 146
Trotlines 52, 56
Trypsin 36
Trypsin-like 11, 13, 18
T-test 100, 136
Tubifex sp. 25, 38, 79, 81–84
Turbidity 58
Tyrosine 93–94

Umpqua River 23
USSR 8, 10, 38, 62, 69, 88, 116, 147–150
Urea extracts 13, 14

Valine 93–94
Variant allele 105–109
Vegetal pole 13, 20, 45
Ventral walls 34
Virginia 62, 64, 70–71
Vision 31, 37
Vital statistics 74–75

Washington 122
'Watch Species' list 136
Water velocities 19, 111–116
Water volume 85
Waterfalls 65
Weaning 38–39, 85
Weight:
–age ratio 84
commercial avg. 67, 120
experiment groups 25
gains on rations 39
hatchery 148–149
lake sturgeon 139–141
–length ratio 67–68, 144
sport avg. 120
tagging 113
white sturgeon 121, 129
Weirs 63
West Virginia 8
White sturgeon:
amino acid levels 93–95
breeding 149
Columbia River 9, 119–125
digestive system 36–41
distribution 6
effect of temperature 43–50
embryos 150
fisheries 147
gamete interaction 11–22
genetic structure 105–109
Idaho 127–133
oocyte maturation 87–92
osmoregulation 23–30
spawning 111
symposium 8
Wild Rose 78–85
Wisconsin 8, 10, 73–78, 135–146

Yellowstone River 51
Yield 123, 149
Yolk 35–37, 47, 149–150
Yolksac 36, 47, 65, 79–83